International Conference on Advanced Materials Behavior and Characterization (ICAMBC 2020)

Held online due to COVID-19

IOP Conference Series: Materials Science and Engineering
Volume 961

Chennai, India
18 - 23 July 2020

ISBN: 978-1-7138-3178-5
ISSN: 1757-8981

Printed from e-media with permission by:

Curran Associates, Inc.
57 Morehouse Lane
Red Hook, NY 12571

Institute of Physics
Dirac House, Temple Back
Bristol BS1 6BE UK

Phone: 44 1 17 929 7481
Fax: 44 1 17 920 0979

techtracking@iop.org

Additional copies of this publication are available from:

Curran Associates, Inc.
57 Morehouse Lane
Red Hook, NY 12571 USA
Phone: 845-758-0400
Fax: 845-758-2633
Email: curran@proceedings.com
Web: www.proceedings.com

TABLE OF CONTENTS

ATOMIC PHYSICS

CIVIL ENGINEERING

MECHANICAL ENGINEERING

PHYSICS

COMPUTATIONAL SCIENCE

CIVIL ENGINEERING

APPLIED CHEMISTRY

CIVIL ENGINEERING

MECHANICAL ENGINEERING

IOP Conf. Series: Materials Science and Engineering **961** (2020) 011001 doi:10.1088/1757-899X/961/1/011001

Preface

The volume of the content has been published through ICAMBC_2020 (International Conference on Advanced Materials Behavior and Characterization_2020) conducted as virtual online conference on 18-23, July 2020. The conference was organized by Mattest Research Academy, Chennai, Tamil Nadu, India.

The focus of the conference is to promote the innovative works from core and diversified fields of materials engineering for different applications. Materials engineering is one of the very most important interdisciplinary fields that focuses on studies fundamental physical and chemical basis in order to develop new materials and compounds through manipulating atoms to form desired structures and properties. Any revolutionary changes are always through the breakthrough in new materials.

The manuscripts were submitted to the conference from multidiscipline of Materials, Engineering and sciences. The authors and participants from international universities and National institutions and Research organizations were joined through online via zoom meet. The experts and session chairs have provided talk related to recent research innovations and importance of the virtual conference in this pandemic situation due to COVID'19.

More than hundreds of manuscripts have been presented in front of 300 hundred participants per day also the event was telecasted in social media like You tube and Facebook.

The forum comprised with ten invited talks in the area of Bio sciences Medical sciences recent advanced materials such as nano materials polymers, ceramics natural fiber reinforced composites, phase change materials, surface engineering, tribology and biomaterials.

Since the conference (ICAMBC_2020) was scheduled on May 18-20, due to COVID'19 it has been rescheduled on June month but considering the restriction from the government of India for social gathering and pandemic it has been organized virtual mode on 18-23, July 2020. The mode of the conference was suggested by the advisory committee, experts and by the authors and participants.

The conference really got a good opening to the research society and everyone encouraged about the event was conducted for six days in pandemic situation. Each presentation is around 20-30 minutes as parallel sessions in front of subject experts. The interaction session was made the real experience to the participants and they had good exposure with the foreign university students and delegates. The global participants from All states in India Iraq, China, South Africa, Nigeria, Iran, Turkey, Morocco Ethiopia, Bangladesh, Canada had

IOP Conf. Series: Materials Science and Engineering **961** (2020) 011001 doi:10.1088/1757-899X/961/1/011001

presented and participated in the conference. Indeed it was a wonderful experience for the global researchers to share their research innovations and the feedback is the evidence from the participants registered in our website. https://www.mattest.net/feed-back

Dr.I.Saravanan
Convener-ICAMBC_2020

Disclaimer

International Conference on Advanced Materials Behavior and Characterization_2020) was conducted as virtual online conference organized by Mattest Research Academy , Chennai, Tamilnadu, India.

Since the conference (ICAMBC_2020) was scheduled on May 18-20, due to COVID'19 it has been rescheduled on June month but considering the restriction from the government of India for social gathering and pandemic it has been organized virtual mode on 18-23, July 2020. The mode of the conference was suggested by the advisory committee, experts and by the authors and participants.

Around 500 participants from all states in India, Iraq, China, South Africa, Nigeria, Iran, Turkey, Morocco Ethiopia, Bangladesh, Canada had presented and participated in the conference through Zoom Meet.

Each presentation is around 20-30 minutes as parallel sessions in front of subject experts. The interaction session was made the real experience to the participants and they had good exposure with the foreign university students and delegates.

Indeed it was a wonderful experience for the global researchers to share their research innovations and the feedback is the evidence from the participants registered in our website. https://www.mattest.net/feed-back

Dr.I.Saravanan
Convener-ICAMBC_2020

IOP Conf. Series: Materials Science and Engineering **961** (2020) 011002 doi:10.1088/1757-899X/961/1/011002

Peer review declaration

All papers published in this volume *of IOP Conference Series: Materials Science and Engineering* have been peer reviewed through processes administered by the Editors. Reviews were conducted by expert referees to the professional and scientific standards expected of a proceedings journal published by IOP Publishing.

- **Type of peer review: Double Blind**
  - **The double-blind peer review process has been followed by sending the manuscripts to the subject experts (Reviewer) without author's name and their details. Authors are informed to incorporate the review comments has been suggested by the reviewers.**
  - **The revision submission of manuscripts has been ensured by the editor (Dr.I.Saravanan) and it has been submitted to the journals through google drive. I assured that if there is any comments and suggestions from the journal, it will be incorporated further.**

- **Conference submission management system: Submission has been received through mail and Peer Review has been followed. • Email ID used for all correspondence and submission is is.icambc@gmail.com (ICAMBC_2020 official Mail ID) by Dr.I.Saravanan, Editor & Convener, ICAMBC_2020**

- **Number of submissions received: 50**

- **Number of submissions sent for review: 50**

- **Number of submissions accepted: 18**

- **Acceptance Rate (Number of Submissions Accepted / Number of Submissions Received X 100): 36 %**

- **Average number of reviews per paper: 3 (1-Preliminary-** Format, Language, Reference Cross Check and Title search + **2-Technical)**

- **Total number of reviewers involved: 30**

- **Contact person for queries:**

Dr.I. Saravanan, Convener & Guest Editor,ICAMBC_2020.
Email : is.mech@outlook.com
Phone : 98414 44418

IOP Conf. Series: Materials Science and Engineering **961** (2020) 012001 doi:10.1088/1757-899X/961/1/012001

Review of Magnesium Metal Matrix Composites

Ravikumar. Saranu[1], Ratnam. Chanamala[2], SrinivasaRao. Putti[3]

[1]Assistant Professor, Department of Mechanical Engineering, Bapatla Engineering College, GBC Road, Mahatmajipuram, Bapatla, Guntur District, Andhra Pradesh-522102, India.
[2]Professor, Department of Mechanical Engineering, Andhra University, Waltair Junction, AU North Campus, Visakhapatnam, Andhra Pradesh-530003, India.
[3]Professor, Department of Mechanical Engineering, Andhra University, Waltair Junction, AU North Campus, Visakhapatnam, Andhra Pradesh-530003, India.

E-mail: ravikumar.saranu@gmail.com

Abstract. In recent years, Magnesium Metal Matrix Composites (Mg-MMC) plays a significant part in an extensive application such as chemical, automotive and aerospace application as because of its lightweight structural and low density properties. In this present investigation, the literature review of Mg-MMCs study includes, reinforcing substantial, processing techniques, Characterization, tribological and mechanical behavior of Mg-MMCs. Furthermore, the CNT, Carbonaceous, Fullerene, SiC, Al_2O_3, TiC, B_4C, Graphenereinforced Mg-MMC articulates are explained in this study. Consequently, the mechanical behavior of tensile strength, hardness, compressive strength and tribological properties of wear rate and corrosion performance was analyzed for various state of the art methods. From the observation, it shows that the ceramic reinforcing increases the hardness and strength of the material. Nevertheless, the ductility was diminished, yet, the ductility and strength of Mg-MMCs can enhance by the titanium based metallic reinforcement.

1. Introduction

In topical centuries, the usage of mg based material is increased due to the aspects of environmental and economic, because mg is a lightweight material contrasted with other iron and aluminum based material [1-2]. Moreover, the mg has $1.74\, gr\, per\, cm^3$ of low density, high rigidity, finest machinability and specific toughness [3]. For this reason, the mg alloys are extensively utilized in electronic industries, aerospace and automotive applications [4]. Nevertheless, the lightweight and more specific toughness materials are require to enhancing the efficiency because the structure of mg is a hexagonal sealed filled construction and less ductility [5-6]. Moreover, the mg has poor wear resistance, average corrosion resistance, and underprivileged mechanical properties [7]. Therefore, mg based alloys are not utilized in profitable utilization compared with the iron and aluminum [8]. In the application of aerospace and automotive, the weightless, finest mechanical characteristics and good wear resistant material characteristics are essential [9-10]. Hence, the tribological behavior of mg and its compounds are ought to be enhanced to satisfy the applications demand [11]. As a result, the researchers have interested to propose tribological behavior of mg alloys by means of mg with composite categorization [12]. The new MMCs material is formed by the combination of matrix along with reinforcement's process [13]. The new MMCs have various merits such as, under high temperature provides finest properties, more specific toughness and less constant of thermal increases [14]. As because of these advantages, the MMCs are the advance composites [15]. While MMCs are designed by the casting technique has less standardized separation of matrix material compared with Powder Metallurgy (PM) method [16]. Nevertheless, the PM process can provide equivalent microstructure without dendritic segments; these may affect the mechanical properties [17]. Besides, the tribological and mechanical properties of Mg-MMCs are improved by the combination of reinforcement particles namely, Al_2O_{3p}, graphene, fiber, SiC, Fullerene, B_4C and etc., [18]. The below sections described the recent techniques of Mg-MMC enhancement and its properties analysis.

2. Reinforcement processing material for synthesis of Mg-MMCs

2.1. *SiC Reinforced Mg-MMC*

Huang, Song-Jeng*et al* [19] investigated a $SiCp/AZ61$ strengthened Mg-MMCs are fabricated using stir casting technique and enhanced the micro plastic distortion by the heat behavior process. Furthermore, the five weight percentage of $SiCp/AZ61$ Mg-MMC fabrications significantly lessen the ductile and shear strength. S. Seshan*et al* [20] studied discontinuously SiC_p reinforced $AZ91$ and $Mg-6Al$ alloy for the fracture behavior and tensile strength, having the size $20\mu m$ .JanuszLelito*et al* [21] proposed SiC strengthened materials are embedded on the grain thickness in an $AZ91$ alloy MMC. The simulation done on this method observed the quantitatively outcomes of cooling turn, the $AZ91/SiC$ fabrication matrix grain thickness in heat separation, SiCmaterial diameter and mass segment. Lim *et al* [22] studied $AZ91$ Mg-MMCs reinforced with SiC for wear performance. Juanjuan*et al* [23] fabricated WS_2 (5 wt%) and SiC (15wt% to 20wt%) reinforced Mg-MMCs using PM technique for enhancing the wear and friction (0.1 to 0.2) behavior. S. Dinesh Kumar *et al* [24] investigated Mg-MMCs reinforced with SiC using PM technique for the improvement of mechanical properties. Enrique *et al* [25] proposed co-continuous ceramic reinforced MMCs based on the combination of metal part and ceramic. The formation of this reinforcement process is done by solid state method. JianghuaShen*et al* [26] studied the 5 to 15 wt% β phase SiC reinforced Mg-MMCs, this method is referred to as MM Nano Composites (MMNC). When the reinforced nanoparticles are in the weight of 10%, then the ultimate strength and yield was enhanced. ParidhiMalhotra*et al* [27] fabricated a new hybrid MMCs by the combination of 7075 Aluminum strengthened with 10 wt% ofSiC and Mg alloy of nanoparticles, which is represented as $Al7075/10\%-SiC/Mg$. Moreover, in the course of EDM rotary working Electrode Wear Rate (EWR) and Material Removal Rate (MRR) are estimated. Kavimani*et al* [28] developed a hybrid composites by different weight percentage of SiC incorporated with reduced Graphene Oxide(r-GO) is strengthened the $AZ31$ of Mg alloy. The performance is tested under sliding distance, velocity and provided load. R. Arrabal*et al* [29] proposed Plasma electrolytic oxidation on $ZC71/12\%(SiCp)-T6$ reinforced MMCs for outstanding performance of specific toughness and low density. From the findings, the 3.4 GPa of hardness attained at process time 100 min and more strength. Karthick*et al* [30] proposed a hybrid form $AZ31$ Mg-MMCs reinforced with SiC and Al_2O_{3p} for enhancing the hardness up to 75.16HV.

2.2. *Al_2O_3 Reinforced Mg-MMC*

Huang and Lin [31] developed a $AM60/Al_2O_{3p}$ reinforced Mg-MMCs for an improvement of mechanical properties using Equal Channel Angular Extrusion (ECAE). Moreover, the stir-casting technique is utilized to fabricate the lightweight of Al_2O_{3p} of 1 wt%, 2 wt% and 5 wt% nano sized particulates aimed at ECAE. Thirugnanasambandham*et al* [32] proposed Mg-MMCs reinforced with

Al_2O_{3p} in 50 nm size using stir casting method for the wear properties. The test case is applied at 10 N and 20N of load with various sliding speed. Song-Jeng Huang *et al* [33] investigated the $AM60/Al_2O_{3p}$ MMCs using ECAE. Moreover, by means of stir casting technique, the current 1wt% of $AM60/Al_2O_{3p}$ particles reinforced Mg-MMCs are explored for ECAE. Tarasasankaet *al* [34] studied the $AZ91E$ mg alloy reinforced with Al_2O_{3p} for optimizing the wear characteristics of MMCs. The analysis are carried out under 1wt % reinforcement, 10N applied load and 500m of sliding distance. Moreover, Niranjanet *al* [35] proposed various weight percentage of Al_2O_3 reinforced $AZ91$-MMCs for the surface integrity. Sameer Kumar *et al* [36] fabricated a nanocomposites of $AZ91E$ with Al_2O_{3p} for enriching the hardness, ductility and wear resistance using stir casting approach. Amandeep Singh and NirajBala [37] introduced Mg Al_2O_{3p} MMCs for the synthesis of wear and hardness performance using dissimilar weight percentages.

2.3. *CNT Reinforced Mg-MMC*

The tribological behavior of mg alloy with the reinforcements of Carbonaceous (C) based accumulation is require to enhance the properties. Therefore, MuhammetEmreTuranet *al* [38] proposed Multi-Walled Carbon Nanotubes (MWCNT) reinforcement and a hybrid compound of MWCNT based Graphenenanoplatelets (GNPs) strengthened material for enhancing the wear behavior of AZ91 mg alloy. Moreover, the proposed composites were produced by the process of burning insistent through semi power metallurgy. Muhammad Rashadmet *al* [39] proposed a new MW-CNTs and GNPs of hybrid combination model, named as CNTs+GNPs for mitigating the demerits of solo GNPs reinforcing. Yunpeng Ding *et al* [40] proposed a new CNT strengthened Mg nanocomposites by means of PM method. From this method, the Yield Strength (YS) is 454 MPa and strength of compression obtained as 504MPa.

2.4. *Fiber Reinforced Mg-MMC*

The strength of the light weight material nanocomposites estimation is essential. Hany S. Abdoet *al* [41] fabricated a combination of particles such as TiO_2 as the nanofiber ceramic material and Carbon Nanofibers (CNF) for investigating the MMCs mechanical characterizations. This work mainly focuses on the MMCs and reinforcement relationship in terms of weight fraction, arrangement and shape. Feng Wu *et al* [42] studied the combination of graphene and carbon fiber material for reinforced Mg-MMCs. Moreover, a huge amount of $Mg_{17}Al_{12}$ particles impulsive in the border sheet in an intermittent manner. Lehua, *et al* [43] proposed carbon fiber reinforced MMCs for strengthening the parameters. Furthermore, Bakkar and Neubert [44] introduced the AS41/ Mg-MMCs with Carbon fiber reinforcement for the minimization of corrosion. The SEM and EDX tests were achieved for the corrosion analysis.

2.5. *TiC reinforced Mg-MMC*

X.Y. Guet *al* [45] fabricated TiC reinforced MMCs ($TiC_p/AZ91D$) with Transient Liquid Phase (TLP) using aluminum interlayer based up on the connection of temperature and time. When the connection temperature is 460^0C at the time increase from 1 minute to 1 hour, then concentration of Al in the junction diminishes. Therefore, the microstructure of ($Al_{12}Mg_{17}$) and TiC is varied, then the quantity of ($Al_{12}Mg_{17}$) is diminished inclination. Anbuchezhiyanet *al* [46] introduced $AZ91D$ Mg nanocomposites with TiC reinforcement under different weight percentage using PM technique, size of 40 μm. Dash *et al* [47] studied AZ91D of mg –MMCs with TiC strengthening for mechanical and tribological performance.

2.6. *Fullerene Reinforced Mg-MMC*
MuhammetEmreTuran [48] is Fabricated a C60 based AZ91 MMCs along with the identical 0.50wt% fraction of weight through SPM and warm pressing method. Moreover, MWCNT, GNPs and C60 reinforcement method is utilized in this approach. MuhammetEmreTuran*et al* [49] investigated fullerene reinforced MMCs, which is prepared using Semi Powder Metallurgy (SPM) using aeration, warm pressing, ultrasonication and sintering. Moreover, from the observation of microstructural the fullerene reinforced MMC is consistently connected. MuhammetEmreTuran*et al* [50] proposed a C60 reinforced Mg-MMC using SPM technique with warm pressing method beneath pure argon environmental condition.

2.7. *B_4C particulate reinforced Mg-MMC*
Q.C. Jiang *et al* [51] fabricated by the dissimilar divisions of 10 wt% to 20 wt% B_4C material reinforced with Mg-MMCs using PM approach. From the results show that the new MMC attained superior wear resistance for 35N at 20 wt % is 5.7039 and hardness 133.3 HB at 20 wt % and 44.0HB at 10 wt%. Therefore, this developed feasible product is more significant and economical for MMCs development. Aydin*et al* [52] studied AZ91 Mg-MCCs reinforced with B_4C under different weight fraction such as 10 to 30 wt% by means of PM technique. Furthermore, wear behavior is validated by various load condition. Therefore, the hardness, YS and UTS are increased.

2.8. *Graphene reinforced Mg-MMC*
The GNPs reinforced Mg-MMCs using semi PM technique introduced by Saberi, *et al* [53]. This method proved as a higher corrosion resistance than ordinary Mg. The antibacterial method increased the concentration of GNPs. Kavimani*et al* [54] proposed AZ31alloy is consistently strengthened with 0.2 wt % to 0.5 wt% of diminished graphene oxide (r-GO). Furthermore, four sequential processes have to be done in this fabrication process namely, solvent treating, automated alloying, cold persistent and sintering below argon condition at $560^0 C$. LinglongMeng*et al* [55] fabricated GNPs reinforced Mg-MMCs for enhancing the toughness and tensile strength. Muhammad Rashad, *et al* [56] developed carbon nanotubes along with graphenenanoplatelets strengthened $AZ31$ Mg alloy by means of PM approach. Therefore, the tensile and compression strength was enhanced due to the reinforcement of carbon nanotubes. The $AZ31$with carbon nanotubes composite was expressively diminished the compressive fracture strain up to negative of 14%. Muhammad Rashad*et al* [57] studied graphenenanoplatelets strengthened $AZ61$ Mg alloy using fragmented thaw deposition technique. Besides, the temperature from $75-225^0 C$ and basic strain value $2\times10^{-3}s^{-1}$ are executed for the study of proposed material tensile strength. Munir*et al* [58] investigated the reinforcement technique of mg-MMCs with different wt % of GNPs developed by PM method for expands the corrosion and mechanical performance. The fabricated composites microstructure and its properties are examined by the various characterization methods.

3. Processing of Metal-Matrix Nanocomposites
The Mg-MMCs developed by the numerous methods of processing technologies.

3.1. *Powder metallurgy processing*
In PM process has four steps have to be done. Primarily, the Mg and the reinforcement materials are crushed in to the powder, and then it is combined, forced and sintered at under the regulated atmospheric temperature [59-61]. Moreover, this method has large possible wt% of elements taken yet this method of process has more. By means of PM method various MMCs such as C60/AZ91, B_4C , GNPs/ $AZ31$, CNT/Mg nanocomposites and Mg-SiC have been fabricated [62]. The schematic illustration of PM steps is illustrated in Fig.1.

3.2. *Stir casting method*

Stir casting method of composite element is a most general technique for fabrication [63]. The graphic illustration of this system set up is illustrated in Fig.2.The reinforcement material such as $AM60/Al_2O_{3p}$, $SiCp/AZ61$ are mixed with MMCs by the assist of stirrer. The MMCs liquid are heated under certain temperature and the reinforcement material is produced. Then, the wettability of reinforcement and matrix stage problems are overcome by scattered in to the molten of matrix using mechanical stirrer [65]. The Molten mixture element is case by conventional casting technique. Nevertheless, the scattering of strengthened composites in to medium is not totally same. Due to the dissimilar densities some groups are formed. In advanced casting method, the MCs are superheated beyond its melting point. Then, the complete melted metal is starts to stirrer [66-67]. In this process, the ultimate solid material is obtained from the melt liquid under the room temperature cooling progress. This method allows 30 wt% of strengthening the matrix stage. Therefore, in MMCs fabrication stir casting has low cost and simple to use. Besides, most of the industries are utilized this technique due to its advantages.

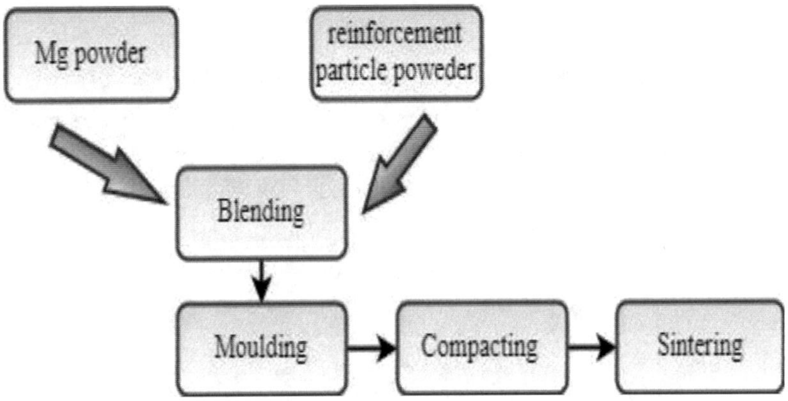

Figure 1. Flow of PM technique process [62].

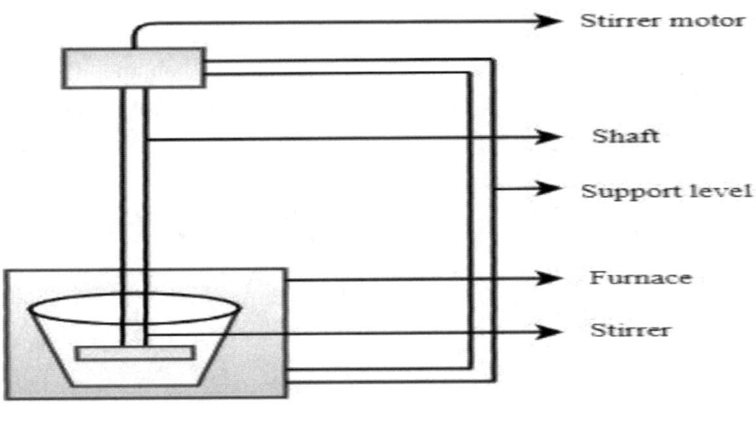

Figure 2. Graphic illustration of Stir casting system [64].

4. Microstructural Characterization

The characterization of the Mg-MMCs is tested using the devices of X-ray Diffraction (XRD) [68], Energy-Dispersive X-Ray Spectroscopy (EDX), Scanning Electron Microscope (SEM) [69] and Field Emission SEM. Moreover, during the analysis of SEM was executed with the accelerating voltage 20

KV, gap 60 μm and 9 mm of processing distance. The composites wear appliances and surface morphology is examined using SEM images [70]. Due to the electron beam the rough interface surface developed, which was examined by SEM and EDX. Consequently, the SEM and EDX are equally distributing the reinforced composite material in MMCs and boost up the relationship of reinforcement and matrix particles.

5. Mechanical Properties

The composites mechanical performance is estimated by the hardness, tensile strength and compressive strength analysis.

5.1. *Hardness*

The hardness of reinforced composites of SiC, CNF, Al_2O_3, TiC, fullerene and B_4C are elaborated in table1. The addition of 0.5wt % of fullerene improved the hardness performance of strengthened Mg-MMC. The C based reinforcement has been enriched the performance of AZ91 alloy in terms of its corrosion rate and hardness [71]. Consequently, the final consecutive strength is improved to 281 MPaatTiO_2 of 5 percentage weight and enhanced the hardness up to 64% by this ceramic based reinforcement. The micro hardness of $AZ61$ Mg alloy for 2wt% of$SiCp/AZ61$ is 76.9HV and 5wt% of $SiCp/AZ61$ is 78.64 HV at the ageing time of thirty six hours [19]. In hard material, the increase of weight portion augmented the MMCs hardness. Consequently, after four passes the 5wt% of $AM60/Al_2O_{3p}$ strengthened MMCs finest hardness is improved to 103.7HV [31]. Moreover, 20 wt% SiC/(r-GO) is strengthened $AZ31$ of Mg alloy hardness is enriched approximately 66 HV. The proposed 0.5 wt% of r-GO fabrication enhanced hardness 64.6HV and at corrosion condition the inhibition of yield efficiency is up to 84% [54]. The AZ91/30wt% B_4C is significantly increased the hardness as 124 HV.

Table 1. Mechanical properties of various composites

S.NO	Ref	Composites	YS (Mpa)	UTS (MPa)	Hardness (HV)	Ductilit y (%)	CS (Mpa)
1	[24]	Mg-4wt% SiC	-	-	40	-	474
		Mg-8wt% SiC			60		524
		Mg-12wt% SiC			70		548
2	[35]	AZ91E+1wt% Al_2O_3	120	160	89	3.08	-
		AZ91E+1wt% Al_2O_3	140	200	99	2.5	
		AZ91E+1wt% Al_2O_3	150	190	82	1.94	
3	[40]	Mg/Ni-CNT	454	504	-	10.5	504
4	[41]	Mg-1wt%/CNF	100	-	46.7	-	221
		Mg-3wt%/CNF	114		44.8		201
		Mg-5wt%/CNF	97		43.9		174
		Mg-10wt%CNF	91		42.5		178
5	[46]	AZ9 1D+3wt%TiC	-	-	55.7	-	35
		AZ9 1D+6wt%TiC			62		34
		AZ9 1D+9wt%TiC			91		45

6	[48]	Mg-Al –0.25wt% fullerene Mg-Al –0.5wt% fullerene	135 178	-	50 54	17.18 15.72	237 296
7	[52]	AZ91/30wt% B_4C	176		124	12	420

5.2. *Tensile strength*

The tensile and yield strength of reinforced composites of SiC, CNF, Al_2O_3, TiC, fullerene and B_4C are elaborated in table1. The CNTs added GNPs reinforced Mg-1Al-0.6wt% of MMC obtained the performance enhancement in terms of its failure strain, yield strength, ultimate tensile strength and elastic modules [72-73]. The developed $AM\,60\,/\,Al_2O_{3p}$ MMCs improves the mechanical properties after four permits of ECAE namely ductility (+125%), YS (+135%) and UTS (+107%) [31]. The shear strength of $(TiC_p\,/\,AZ91D)$ method attained 58 MPa at the bonding temperature 460^0C and time $20\,\text{min}$ [45]. Consequently, the GNPs composites attained 160 MPa of tensile strength at 0.25 wt% and 170 MPa of tensile strength at 0.75 wt%. Nevertheless, the pure Mg has attained 136 MPa of tensile strength [56].

5.3. *Compressive strength*

The compressive of reinforced composites of SiC, CNF, Al_2O_3, TiC, fullerene and B_4C are elaborated in table.1. As per the standard of ASTM B783, the solid Mg composites Compressive Strength (CS) were estimated using universal test equipment [74]. By means of Mg-MMCs with 4 wt%, 8wt% and 12 wt% of SiC attained improved compressive strength. The highest CS 548MPa is obtained at the 12wt% of SiC in Mg alloy [24]. The CS of various composites under different weight fractions is examined. Besides, the Mg-4wt% SiC are significantly increased the CS as 474MPa.

6. Tribologicalbehavior

6.1. *Wear performance*

In MMCs modeling, mainly three methods have been utilized for the estimation of wear behavior such as tilling, crossing point and removal of entire materials [75, 76]. Moreover, the wear rate is impacted by the reinforcing material dimension and distribution. However, the tilling process goes to dominant, if the saturation depth using rough pin is higher than the dimension of reinforcement [77-79]. The wear rate of various composites under applied load condition is shown in Fig.3.

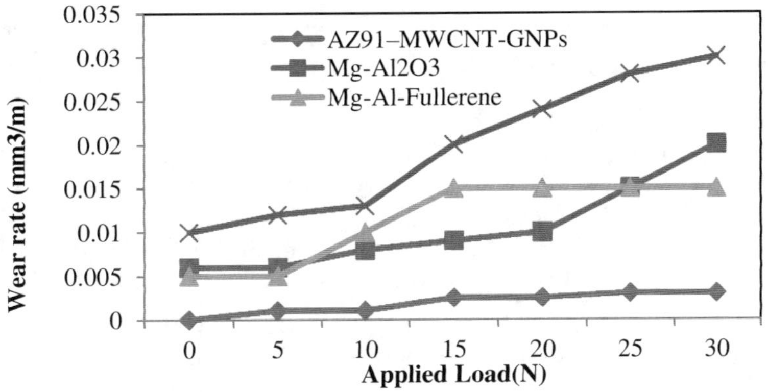

Figure 3. Wear rate under applied load of Mg-MCCs fabrication [38], [32], [48], [52].

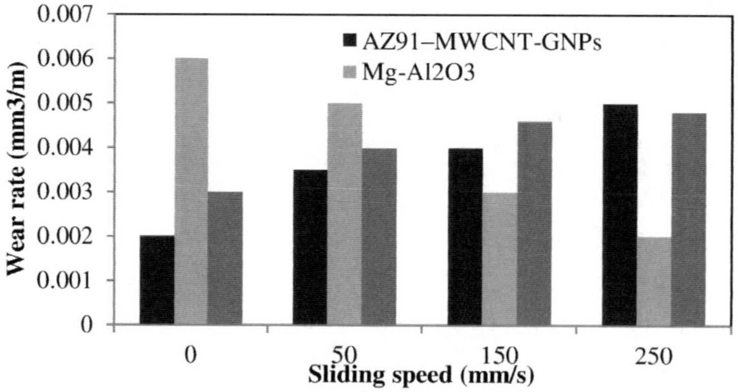

Figure 4. Wear rate under sliding speed of Mg-MCCs fabrication [38], [32], [28].

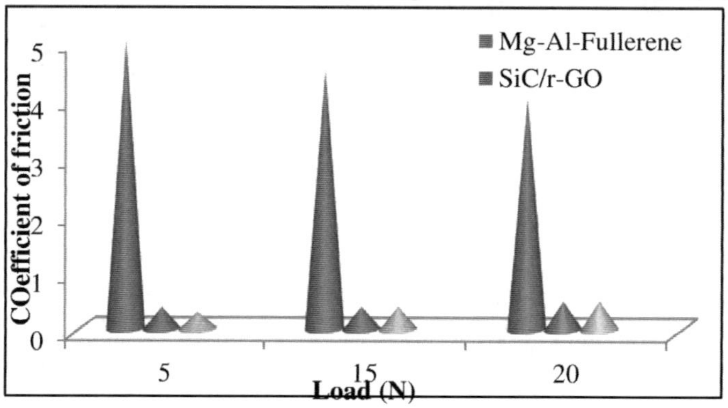

Figure 5. CoF of Mg-MCCs fabrication [48], [28], [32].

The equivalent carbon atom distribution is attained on both 0.15 wt% of AZ91in MWCNT and GNPs composites, thus it provided finest performance of wear resistant. The addition of 0.5wt % of fullerene improved the wear and hardness performance of strengthened Mg-MMC. The wear rate is diminished up to 64% using the ceramic reinforced MMCs. The proposed 0.5 wt% of r-GO fabrication attained decreased wear rate in the range of $2.6mm^3Nm^{-1}$.The wear rate of different composites under sliding speed is illustrated in Fig.4.The addition of $Al_2O_{3_p}$ in MMCs decreased the wear rate 183 μmunder typical 10 N load states with 800rpm sliding speed [32]. The Coefficient of Friction (CoF) for different composites is illustrated in Fig.5.

6.2. *Corrosion rate*

The corrosion rate of Mg-MMCs [80] was validated as per the standard of ASTM B-117 in a shower of salt at forty eight hours [81, 82]. From the observations, the AZ91D Mg alloy corrosion value is mainly based on the primary Mg composition, subsequently, it attained higher corrosion value than the secondary Mg state [83]. The C based reinforcement has been enriched the performance of AZ91 alloy in terms of its hardness. The $ZC71/12\%(SiCp)-T6$ reinforced MMCs has the 3.5 wt% of $NaCl$ is utilized in corrosion resistance rate reduction and hardness of coatings are evaluated. Due to the current density of corrosion increase, the electron flow was also increases [84], hence the corrosion rate was increased.

7. Application

The vehicle usage is increased in recent year; therefore the lightweight Mg based alloys are utilized in automotive application. Initially, the wheels of the vehicles are alternated by Mg alloy [85-86]. The interior, engine and transmission sections, body of tools, and chassis are developed by the lightweight fabricated Mg alloys [87]. The emission of CO_2 is diminished and reduced the weight due to the increase of Mg alloys in automotive industries [88]. In medical application, the bone infection injuries treatment is replaced by the Mg-GNP composites. Pedersen and Ramulu*et al* [89] introduced the carbide cutting tools for industries. The binder jet additive are designed effectively by the new CNT reinforced Mg-MMCs composites [90]. Thus the Mg-MMCs play an essential role in aerospace, glass, wheel and automotive industries.

8. Conclusion

In this study, the various reinforcement materials, tribological and mechanical performance of MMCs are examined. In order to improve the engineering application of Mg alloy, various kinds of MMCs are developed by dissimilar dimensions, configuration and reinforcement elements category. The proper fabrication and reinforcement of MMCs are the main factor for the composition of microstructure and mechanical performance based on the necessities. From this investigation of review was observed follows, The GNP reinforcement method is a new one, because it has attractive mechanical properties. Nevertheless, the unique attractive characterization of GNP mechanical properties is transfer to combine. This is restricting the GNP effectiveness the application of MMCs. The Al_2O_3 and SiC reinforced Mg alloy based MMT provides finest relationship to the fiber matrix contrast to CNTs. Thus the reinforced MMCs attained more influence stress, more tensile strength, less thermal growth and more hardness. Besides, the CNTs reinforced MMCs attained high strength and low mass density due to the friction coefficient reduction. While compare to the Al_2O_3 and SiC, the CNT fabrication provides light weigh MMCs. The corrosion resistance properties are reduced using SiC, yet the properties are enhanced by other reinforced particles. Besides, the mechanical and tribological features are enhanced using new organic material such as fly ash etc., based hybrid composites.

References

[1] Kim W J, An C W, Kim Y S and Hong S I 2002 Mechanical properties and microstructures of an AZ61 Mg Alloy produced by equal channel angular pressing *Scriptamaterialia* **47**(1) 39-44
[2] Park S H C, Sato Y SandKokawa H 2003 Effect of micro-texture on fracture location in friction stir weld of Mg alloy AZ61 during tensile test *ScriptaMaterialia* **49**(2) 161-166
[3] Pradhan M K 2020 TribologicalBehavior, Machinability, and Optimization of EDM of AA-2014 Hybrid Composite Reinforced With SiC and Glass Particulates *Handbook of Research on Developments and Trends in Industrial and Materials Engineering* (IGI Global) pp228-269
[4] Romanowski C 2019 Magnesium Alloy Sheet for Transportation Applications *Magnesium Technology 2019* (Springer: Cham) pp3-12
[5] Kumar D SandSasanka C T 2017 Magnesium and Its Alloys *Lightweight and Sustainable Materials for Automotive Applications* (CRC Press) pp329-368
[6] Friedrich H EandMordike B L 2006 Engineering requirements, strategies and examples *Magnesium Technology: Metallurgy, Design Data, Applications* 499-632
[7] Majumdar J D, Galun R, Mordike B Land Manna I 2003 Effect of laser surface melting on corrosion and wear resistance of a commercial magnesium alloy *Materials Science and Engineering: A* **361**(1-2) 119-129
[8] Wallace F J 1963 The determination of magnesium in aluminium alloys by atomic-absorption spectroscopy *Analyst* **88**(1045) 259-265
[9] Vlasblom M 2018 The manufacture, properties, and applications of high-strength, high-modulus

polyethylene fibers *Handbook of Properties of Textile and Technical Fibres* (Woodhead Publishing) pp699-755

[10]Mohanavel V 2019 Synthesis and evaluation on mechanical properties of LM4/AlN alloy based composites *Energy Sources, Part A: Recovery, Utilization, and Environmental Effects* 1-10

[11]Chan C H, Tang S W, Mohd N Kand Lim W H 2018 Tribologicalbehavior of biolubricant base stocks and additives *Renewable and Sustainable Energy Reviews* **93** 145-157

[12]Gulzar M, Masjuki H H, Kalam M AandVarman M 2016 Tribological performance of nanoparticles as lubricating oil additives *Journal of Nanoparticle Research* **18**(8) 223

[13]Singh S, Singh IandDvivedi A 2017 Design and development of novel cost effective casting route for production of metal matrix composites (MMCs) *International Journal of Cast Metals Research* **30**(6) 356-364

[14]Miracle D B 2005 Metal matrix composites–from science to technological significance *Composites science and technology* **65**(15-16) 2526-2540

[15]Allison J Eand Cole G S 1993 Metal-matrix composites in the automotive industry: opportunities and challenges *JoM* **45**(1) 19-24

[16]Ibrahim M A, Sahin Y, Gidado A Yand Said M T 2019 Mechanical Properties of Aluminium Matrix Composite Including SiC/Al2O3 by Powder Metallurgy-A Review *GSJ* **7**(3) 23-38

[17]Vogiatzis C AandSkolianos S M 2016 On the sintering mechanisms and microstructure of aluminium–ceramic cenospheres syntactic foams produced by powder metallurgy route *Composites Part A: Applied Science and Manufacturing* **82** 8-19

[18]Malaki M, Xu W, Kasar A K, Menezes P LandDieringa H 2019 Advanced metal matrix nanocomposites *Metals* **9**(3) 330

[19]Huang S Jand Ali A N 2018 Effects of heat treatment on the microstructure and microplastic deformation behavior of SiC particles reinforced AZ61 magnesium metal matrix composite *Materials Science and Engineering: A* **711** 670-682

[20]Seshan S, Jayamathy Mand Kailas S V 2003 The tensile behavior of two magnesium alloys reinforced with silicon carbide particulates *Materials science and engineering: A* **363**(1-2) 345-351

[21]Lelito J, Zak P L, Shirzadi A A, Greer A LandKrajewski W K 2012 Effect of SiC reinforcement particles on the grain density in a magnesium-based metal–matrix composite: Modelling and experiment *ActaMaterialia* **60**(6-7) 2950-2958

[22]Lim C Y H, Lim S Cand Gupta M 2003 Wear behaviour of SiCp-reinforced magnesium matrix composites *Wear* **255**(1-6) 629-637

[23]Zhu J, Qi J, Guan D, Ma Land Dwyer-Joyce R 2020Tribological behaviour of self-lubricating Mg matrix composites reinforced with silicon carbide and tungsten disulfide *Tribology International* **146**106253

[24]Kumar S D, Ravichandran MandSakthivelu S 2020 Mechanical properties of magnesium-silicon carbide composite fabricated through powder metallurgy route *Materials Today: Proceedings*

[25]Enrique P D, Marzbanrad EandMahmoodkhani Y 2020 Design of binder jet additive manufactured co-continuous ceramic-reinforced metal matrix composites *Journal of Materials Science & Technology*

[26]Shen J, Yin W, Wei Q, Li Y, Liu Jand An L 2013 Effect of ceramic nanoparticle reinforcements on the quasistatic and dynamic mechanical properties of magnesium-based metal matrix composites *Journal of Materials Research* **28**(13) 1835-1852

[27]Malhotra P, Tyagi R K, Singh N KandSikarwar B S 2020 Experimental investigation and effects of process parameters on EDM of Al7075/SiC composite reinforced with magnesium particles *Materials Today: Proceedings* **21** 1496-1501

[28]Kavimani V, Prakash K S, Starvin M SandKalidas B 2020 Tribo-surface characteristics and wear behaviour of SiC@ r-GO/Mg composite worn under varying control factor *Silicon* **12**(1) 29-39

[29]Arrabal R, Matykina E, Skeldon Pand Thompson G E 2009 Coating formation by plasma electrolytic oxidation on ZC71/SiC/12p-T6 magnesium metal matrix composite *Applied surface science* **255**(9) 5071-5078

[30]Karthick E, Mathai Jand Tony J M 2017 Processing, Microstructure and Mechanical Properties of Al2O3 and SiC Reinforced Magnesium Metal Matrix Hybrid Composites *Materials Today: Proceedings* **4**(6) 6750-6756

[31]Huang S Jand Lin P C 2013 Grain refinement of AM60/Al2O3p magnesium metal-matrix composites processed by ECAE *Kovove Mater* **51**(6) 357-366

[32]Thirugnanasambandham TandChandradass J 2019 Experimental study of wear characteristics of Al2O3 reinforced magnesium based metal matrix composites *Materials Today: Proceedings* **14** 211-218

[33]Huang S J, Lin P CandAoh J N 2015 Mechanical behavior enhancement of AM60/Al_2O_3p magnesium metal–matrix nanocomposites by ECAE *Materials and Manufacturing Processes* **30**(10) 1272-1277

[34]Tarasasanka C, Snehita KandRavindra K 2019 Optimization of dry sliding wear properties ofAZ91E/nano Al2O3 reinforced metal matrix composite with grey relational analysis *International Journal of Engineering, Science and Technology* **11**(4) 41-48

[35]Niranjan C A, Srinivas SandRamachandra M 2020 Experimental investigations on depth of penetration and surface integrity in AZ91/Al2O3 nano-composites cut by abrasive water jet *The International Journal of Advanced Manufacturing Technology* **107**(1) 747-762

[36]Kumar S, Suman K N S, Ravindra KandPoddar P 2017 Microstructure, mechanical response and fractography of AZ91E/Al2O3 (p) nano composite fabricated by semi solid stir casting method *Journal of magnesium and alloys* **5**(1) 48-55

[37]Singh AandBala N 2019 Synthesis and comparative sliding wear behavior of stir cast Mg and Mg/Al2O3 metal matrix composites *Materials Research Express* **6**(7) 076512

[38]Turan M E, Zengin Hand Sun Y 2019 Dry sliding wear behavior of (MWCNT+ GNPs) reinforced AZ91 magnesium matrix hybrid composites *Metals and Materials International* 1-10

[39]Rashad M, Pan F, Tang A, Asif MandAamir M 2014 Synergetic effect of graphenenanoplatelets (GNPs) and multi-walled carbon nanotube (MW-CNTs) on mechanical properties of pure magnesium *Journal of Alloys and Compounds* **603** 111-118

[40]Ding Y, Xu J, Hu J, Gao Q, Guo Xand Zhang R 2020. High performance carbon nanotube-reinforced magnesium nanocomposite *Materials Science and Engineering: A* **771** 138575

[41]Abdo H S, Khalil K A, El-Rayes M MandMarzouk W W 2019 Ceramic nanofibers versus carbon nanofibers as a reinforcement for magnesium metal matrix to improve the mechanical properties *Journal of King Saud University-Engineering Sciences*

[42]Wu Fand Zhu J 1997 Morphology of second-phase precipitates in carbon-fiber-and graphite-fiber-reinforced magnesium-based metal-matrix composites *Composites science and technology* **57**(6) 661-667

[43]Qi L, Li S, Zhang T, Zhou Jand Li H 2019 An analysis of the factors affecting strengthening in

carbon fiber reinforced magnesium composites *Composite Structures* **209** 328-336
[44]Bakkar AandNeubert V 2009 Corrosion behaviour of carbon fibres/magnesium metal matrix composite and electrochemical response of its constituents *Electrochimicaacta* **54**(5) 1597-1606
[45]Gu X Y, Sun D Qand Liu L 2008 Transient liquid phase bonding of TiC reinforced magnesium metal matrix composites (TiCP/AZ91D) using aluminum interlayer *Materials Science and Engineering: A* **487**(1-2) 86-92
[46]Anbuchezhiyan G, Mohan BandKathiresan S 2020 Influence of microstructure and mechanical properties of TiC reinforced magnesium nano composites *Materials Today: Proceedings*
[47]Dash D, Singh R, Samanta SandRai R N 2020 Influence of TiC on Microstructure, Mechanical and Wear Properties of Magnesium alloy (AZ91D) Matrix Composites.
[48]Turan M E, Sun Y, Aydin F, Zengin HandTuren Y 2018 Effects of carbonaceous reinforcements on microstructure and corrosion properties of magnesium matrix composites *Materials Chemistry and Physics* **218** 182-188
[49]Turan M E, Sun YandAkgul Y 2018 Mechanical, tribological and corrosion properties of fullerene reinforced magnesium matrix composites fabricated by semi powder metallurgy *Journal of Alloys and Compounds* **740** 1149-1158
[50]Turan M E, Sun YandAkgul Y 2018 Improved wear properties of magnesium matrix composite with the addition of fullerene using semi powder metallurgy *Fullerenes, Nanotubes and Carbon Nanostructures* **26**(2) 130-136
[51]Jiang Q C, Wang H Y, Ma B X, Wang Yand Zhao F 2005 Fabrication of B4C particulate reinforced magnesium matrix composite by powder metallurgy *Journal of Alloys and Compounds* **386**(1-2) 177-181
[52]Aydin F, Sun YandTuran M E 2019 Investigation of microstructure, mechanical and wear behaviour of B4C particulate reinforced AZ91 matrix composites by powder metallurgy
[53]Saberi A, Bakhsheshi-Rad H RandKaramian E 2020 Magnesium-graphenenano-platelet composites: Corrosion behavior, mechanical and biological properties *Journal of Alloys and Compounds* **821** 153379
[54]Kavimani V, Prakash K SandThankachan T 2019 Investigation of graphene-reinforced magnesium metal matrix composites processed through a solvent-based powder metallurgy route *Bulletin of Materials Science* **42**(1) 39
[55]Meng L, Hu X, Wang X, Zhang C, Shi Hand Xiang Y 2018Graphenenanoplatelets reinforced Mg matrix composite with enhanced mechanical properties by structure construction *Materials Science and Engineering: A* **733**414-418
[56]Rashad M, Pan F, Zhang JandAsif M2015 Use of high energy ball milling to study the role of graphenenanoplatelets and carbon nanotubes reinforced magnesium alloy *Journal of Alloys and Compounds* **646**223-232
[57]Rashad M, Pan F, Lin DandAsif M2016 High temperature mechanical behavior of AZ61 magnesium alloy reinforced with graphenenanoplatelets *Materials & Design* **89**1242-1250
[58]Munir K, Wen Cand Li Y2020 Graphenenanoplatelets-reinforced magnesium metal matrix nanocomposites with superior mechanical and corrosion performance for biomedical applications *Journal of Magnesium and Alloys*
[59]Lloyd DJ1994 Particle reinforced aluminium and magnesium matrix composites *International materials reviews* **39**(1)1-23
[60]Tun KSand Gupta M2007 Improving mechanical properties of magnesium using nano-yttria

reinforcement and microwave assisted powder metallurgy method *Composites Science and Technology* **67**(13)2657-2664
[61]Kumar N, Bharti AandSaxena KK2020 A re-analysis of effect of various process parameters on the mechanical properties of Mg based MMCs fabricated by powder metallurgy technique *Materials Today: Proceedings*
[62]Manige ST, Gowd GHand Chandra Mohan Reddy B2019 Processing of magnesium metal matrix composite reinforced with graphenenano platelets through vacuum stir casting and investigating its mechanical behaviour *Journal of Thin Films, Coating Science Technology and Application* **5**(3)21-29
[63]Venkatesh B, Sandeep Pand Ramakrishna MVA2019Synthesis and mechanical characterization of magnesium reinforced with SiC composites *Materials Today: Proceedings* **19**792-797
[64]Ceschini L, Dahle A, Gupta MandJarfors AEW2017 Ex Situ Production Routes for Metal Matrix Nanocomposites *Aluminum and Magnesium Metal Matrix Nanocomposites* (Springer: Singapore) pp 19-40
[65]Hashim J, Looney LandHashmi MSJ1999 Metal matrix composites: production by the stir casting method *Journal of materials processing technology* **92**1-7
[66]Lan J, Yang Yand Li X2004 Microstructure and microhardness of SiC nanoparticles reinforced magnesium composites fabricated by ultrasonic method *Materials Science and Engineering: A* **386**(1-2)284-290
[67]Parthiban GT, Malarkodi DandPalaniswamy N2010 Corrosion protection by acrylamide treatment for magnesium alloy metal matrix composite (MMC) reinforced with titanium boride *Surface engineering* **26**(5)378-384
[68]Yong MSand Clegg AJ2005 Process optimisation for a squeeze cast magnesium alloy metal matrix composite *Journal of Materials Processing Technology* **168**(2)262-269
[69]Vinoth R, Asharudeen A, Balachandar Eand Kumar D2019 Enhancement of Mechanical properties of Magnesium metal matrix composites
[70]Zhang J, Wei Qand Joshi SP2016 Effects of reinforcement morphology on the mechanical behavior of magnesium metal matrix composites based on crystal plasticity modeling *Mechanics of Materials* **95**1-14
[71]Luo A1995 Processing, microstructure, and mechanical behavior of cast magnesium metal matrix composites *Metallurgical and Materials Transactions A* **26**(9)2445-2455
[72]Purazrang K, Kainer KUandMordike BL1991 Fracture toughness behaviour of a magnesium alloy metal-matrix composite produced by the infiltration technique *Composites* **22**(6)456-462
[73]Wang HY, Jiang QC, Wang Y, Ma BXand Zhao F2004 Fabrication of TiB2 particulate reinforced magnesium matrix composites by powder metallurgy *Materials letters* **58**(27-28)3509-3513
[74]Ferguson JB, Sheykh-Jaberi Fand Kim CS2012 On the strength and strain to failure in particle-reinforced magnesium metal-matrix nanocomposites (Mg MMNCs) *Materials Science and Engineering: A* **558**193-204
[75]Kavimani V, Prakash KSandThankachan T2019 Experimental investigations on wear and friction behaviour of SiC@ r-GO reinforced Mg matrix composites produced through solvent-based powder metallurgy *Composites Part B: Engineering* **162**508-521
[76]Wang N, Wang Zand Weatherly GC1992 Formation of magnesium aluminate (spinel) in cast SiC particulate-reinforced Al (A356) metal matrix composites *Metallurgical Transactions A* **23**(5)1423-1430

[77]Hartmann O, Kemnitzer MandBiermann H2002 Influence of reinforcement morphology and matrix strength of metal–matrix composites on the cyclic deformation and fatigue behaviour *International journal of fatigue* **24**(2-4)215-221

[78]Kavimani V, Prakash KS, Thankachan TandNagaraja S2019 WEDM Parameter Optimization for Silicon@ r-GO/Magneisum Composite Using Taguchi Based GRA Coupled PCA *Silicon* 1-15

[79]Bakkar AandNeubert V2007 Corrosion characterisation of alumina–magnesium metal matrix composites *Corrosion science* **49**(3)1110-1130

[80]Fathi MH, Meratian MandRazavi M2011 Novel magnesium-nanofluorapatite metal matrix nanocomposite with improved biodegradation behavior *Journal of biomedical nanotechnology* **7**(3)441-445

[81]Hihara LHandKondepudi PK1993 The galvanic corrosion of SiC monofilament/ZE41 Mg metal-matrix composite in 0.5 M NaNO3 *Corrosion science* **34**(11)1761-1772

[82]Merino MC, Pardo A, Arrabal R, Merino SandCasajús P2010 Influence of chloride ion concentration and temperature on the corrosion of Mg–Al alloys in salt fog *Corrosion science* **52**(5)1696-1704

[83]Martin HJ, Horstemeyer MFand Wang PT2010 Comparison of corrosion pitting under immersion and salt-spray environments on an as-cast AE44 magnesium alloy *Corrosion science* **52**(11)3624-3638

[84]Skar JIand Albright D2016 Emerging trends in corrosion protection of magnesium die-castings *Essential Readings in Magnesium Technology*(Springer: Cham)pp 585-591

[85]Witte F, Hort N, Vogt C, Cohen SandKainer KU2007 Biodegradable magnesium–hydroxyapatite metal matrix composites *Biomaterials* **28**(13)2163-2174

[86]Luo AA2013 Magnesium casting technology for structural applications *journal of Magnesium and Alloys* **1**(1)2-22

[87]Merlo AM2003 The contribution of surface engineering to the product performance in the automotive industry *Surface and Coatings Technology* **174** 21-26

[88]Watarai H2006 Trend of Research and Development for Magnesium Alloys-Reducing the Weight of Structural Materialsin Motor Vehicles *NISTEP Science & Technology Foresight Center*

[89]Pedersen WandRamulu M2006 Facing SiCp/Mg metal matrix composites with carbide tools *Journal of materials processing technology* **172**(3)417-423

[90]Rajak DK, Wagh PHandMenezes PL2020 Critical Overview of Coatings Technology for Metal Matrix Composites *Journal of Bio-and Tribo-Corrosion* **6**(1)12

Acknowledgments
None

IOP Conf. Series: Materials Science and Engineering **961** (2020) 012002 doi:10.1088/1757-899X/961/1/012002

Experimental investigation on the strength of chipped rubber-based concrete

S Dhiman[1], R Garg[2] , R Garg[3,*], and S Singla[4]

[1]Department of Civil Engineering, CGC Technical Campus, Jhanjeri, Mohali, India
[2]Department of Civil Engineering, Galgotias College of Engineering and Technology, Greater Noida, India
[3]Department of Chemistry, Rayat Bahra University, Mohali, India
[4]Department of Civil Engineering, RIMT University, Mandi Gobindgarh, India
*Corresponding author: rajni.garg@rayatbahrauniversity.edu.in

Abstract. In this research work, the strength and microstructure of chipped rubber-based compressed concrete have been investigated. Chipped rubber has been utilized as an option over coarse aggregate. Chipped rubber at normal instance decreases the durable properties of concrete but the combination of compression testing as well as chipped rubber replacement can significantly affect the strength of the material. During the study, coarse aggregate has been replaced at 0-50% with chipped rubber. The freshly prepared concrete has been condensed in a specially designed mould and after that compression casting technique has been applied. Samples of both compressed chipped rubber as well as uncompressed chipped rubber have been prepared and then examined further for compressive strength, split tensile strength, and abrasion resistance. Microstructures of conventional concrete, compressed rubber concrete, and uncompressed rubber concrete, has also been analyzed. The results show that the concrete with replacement by 20% of chipped rubber gives the significant results for compressive strength and split tensile strength. This proposed method should be implemented by the concrete industry from an environmental point of view.

1. Introduction

In the whole world, approximately 1 billion tires are put in the waste category and this number is increasing day by day [1]. During the recent two to three decades, the automobile industry in the entire world is growing very rapidly due to which the availability of waste tire rubber is also increased. This waste is very much harmful from the environmental point of concern, so strict actions should be taken to dump it off safely and effectively [2]. From the total tire waste, approximately half is burnt which releases harmful gases that further lead harm to the natural parameters of the earth. The rest of this tire waste is dumped into the soil which can further result in drastic environmental issues [3]. Therefore, it's a time clicking for the researchers to find alternative use of this kind of waste in this modern period. Nowadays, some of this waste is also used for various engineering projects but it should be increased as it has very much good effect in the civil industry. This increase will decrease the load of dumping of this waste and will decrease the environmental loss [4,5].

Waste rubber is used in the form of chipped rubber in the construction works [6]. Chipped rubber has some vulnerable properties which can be harnessed to produce concrete structures including beams, columns, sections such as I sections, h section, channel section. Chipped rubber is mainly used as a substituent of natural coarse aggregate and then it is compressed to produce compressed rubber concrete. Chipped rubber is generally replaced from 0 to 50 % for both compressed as well as uncompressed members. The effects of chipped tire rubber on the properties of concrete by replacing the fine as well as coarse aggregate with waste rubber has been reported in the literature. The fine aggregate particles were replaced partially with waste rubber and the results were quite satisfactory as the strength was enhanced at the lower replacement concentrations, but at higher concentration, the strength parameters were adversely affected [7]. Reduction in compressive strength is generally unavoidable when various forms of rubber particles are used for replacing the aggregates partially, irrespective of the particle size, the degree of rubber substitution, or the form of aggregate substituted. In general, a reduced compressive strength was indicated in many cases and a decrease in compressive strength was more conspicuous with an increased substitution level of rubber. This lessening effect is due to poor interfacial interaction between rubber and binder particles and the presence of the entrapped air in the voids at higher replacement levels [8]. Likewise, an increase in water absorption by immersion has been observed on the replacement of natural aggregates by rubber waste. The results were attributed to the requirement of higher water content by concrete with added rubber as well as a higher amount of pores due to poor binding of the rubber and the cement particles [4]. This factor also reduces the density of rubber-based concrete significantly as compared to conventional concrete [9].

The abrasion resistance of rubber concrete has been examined by replacing the fine aggregates with crumb rubber at 0-20% of replacement rates. They noticed that the introduction of crumb rubber increased concrete resistance in an aggressive environment at all replacement rates [10]. The microstructure as well as the inner structure of natural aggregate concrete as well as the waste rubber-based concrete has been compared by SEM analysis. It has been observed that replacing natural aggregate by waste tire rubber results in a denser microstructure as compared to natural aggregate concrete. However, with the increase in the replacement percentage, the microstructure of the matrix gets less dense and thus, strength is affected adversely [11]. The declining compressive strength with the increasing content of embedded rubber can be lessened by replacing a proper amount of cement with strength-enhancing additives like silica fume or by treatment of rubber with sodium hydroxide [12]. The strength and durability aspects of concrete have also been investigated by replacing natural river sand by tire rubber waste with the addition of silica fume [13]. The results specified that although the compressive strength of concrete decreases with the introduction of rubber waste, yet the introduction of silica fume compensates for the degrees of compressive strength [14]. Similarly, the compressive strength of concrete with substitution of aggregate by crumbed rubber in the presence of steel fibers has also been investigated. It has been found that the presence of steel fibers maintains the compressive strength of the concrete that would have been lesser in absence of steel fibers along with the introduction of waste rubber [15].

This study has been aimed to evaluate the strength parameters of concrete by partially replacing coarse aggregate by Chipped Rubber under specially designed compression casting technique and to compare the abrasion resistance for conventional concrete, compressed chipped rubber concrete, and uncompressed chipped rubber concrete. The microstructure of conventional concrete, compressed rubber concrete, and uncompressed rubber concrete has been analyzed using SEM analysis and the results have been correlated with the strength analysis.

2. Materials and Methods

Chipped rubber was collected from the nearest tire recycling plant in Chandigarh (India). The size of the chipped rubber taken for the research work lied in between 10 mm to 20 mm. 43 grade OPC (Ordinary Portland cement) was used for the research work. Coarse aggregates with particle size ranging between 10-20 mm and natural river sand as the fine aggregate was used for the preparation of concrete mix. Table 1 shows the physical characteristics of the aggregates. Three types of mix proportions were prepared with coarse aggregate for conventional concrete (M 30 grade), chipped rubber-based aggregate for uncompressed concrete, and compressed concrete to obtain eleven mixes in total. In the case of rubber aggregate concrete, chipped rubber was used as a replacement of natural coarse aggregate from 0% to 50% by volume. All the Mix proportions have been listed in Table 2. Firstly, the materials were mixed using a mixer as per their quantities. The mixtures for conventional concrete were transferred into the normal mold while the mixes for compressed rubber concrete specimens were transferred into specially designed molds attached with a compressive load. When the load was applied over the molds, the specimens were compressed with the removal of moisture and pores. This kind of compression leads to the condensation of the specimens. The load was removed after 24 hrs. and the samples were subjected to curing for 28 days after demoulding. The compressive and split tensile strength of the specimens was tested as per IS 516 while the abrasion resistance was tested as per IS 1237 [16].

Table 1. Physical Characteristics of Aggregates

Aggregate Type	Bulk Density (kg/m^2)	Water absorption (%)	Specific gravity
Natural Coarse Aggregate	1510	1.25	2.60
Chipped Rubber	695	1.65	1.05
Natural fine aggregate	1615	1.15	2.55

Table 2. Mix proportion of Specimens

Type	Cement (kg/m^3)	Sand (g)	Coarse Aggregate (g)	Chipped Rubber (g)	Water (g)
CC	433.3	701.2	1110.6	-	0.45
URC10	433.3	701.2	999.54	111.06	0.45
CRC10	433.3	701.2	999.54	111.06	0.45
URC20	433.3	701.2	888.48	222.33	0.45
CRC20	433.3	701.2	888.48	222.33	0.45
URC30	433.3	701.2	777.42	333.18	0.45
CRC30	433.3	701.2	777.42	333.18	0.45
URC40	433.3	701.2	666.36	444.24	0.45
CRC40	433.3	701.2	666.36	444.24	0.45
URC50	433.3	701.2	555.30	555.30	0.45
CRC50	433.3	701.2	555.30	555.30	0.45

*Type CC denotes conventional concrete, URC denotes uncompressed rubber concrete and CRC denotes compressed concrete.

3. Results and Discussions

3.1. Compressive Strength

Figure 1 illustrates the effect of the coarse aggregate replacement by rubber waste on the compressive strength of the compressed and uncompressed concrete at 28 days. The replacement by rubber waste in uncompressed concrete was found to result in a substantial strength reduction [17]. The compressive strength of the URC50 concrete was 50.0% lower compared to the CC concrete. For compressed concrete, the increase in strength was noticed up to a substituent dosage of 20% followed by a decrease in compressive strength, w.r.t. conventional concrete. The compressive strength of the CRC20 concrete was 24.0% higher than CC concrete while that of CRC40 concrete was 3.0% lesser than that of the CC concrete, at 28 days (Figure 2). Due to the compression casting mechanism, the pores inside the concrete microstructure get filled leading to the condensation of concrete specimens. This condensation as well as pore pressure densifies the microstructure of the concrete which further enhances the overall strength of the concrete. In general, the substitution by a higher amount of rubber waste (30%) decreased the compressive strength w.r.t. conventional concrete, regardless of the compression treatment performed on the mixes. Rubber is a soft and elastic material in contrast to the other binding materials present in concrete. Hence, there is poor bonding between chipped rubber waste and other binder particles in the concrete matrix. It results in the appearance of micro-cracks in the cement matrix leading to a lessening in compressive strength. The results are consistent with earlier studies reported in the literature [16,18]. The compressive strength of compressed rubber concrete was found better than uncompressed rubber concrete at all replacement amount. The compressive strength of URC10 was noticed as similar to that of CRC50 indicating the positive impact of compression technique on the compressive strength [18].

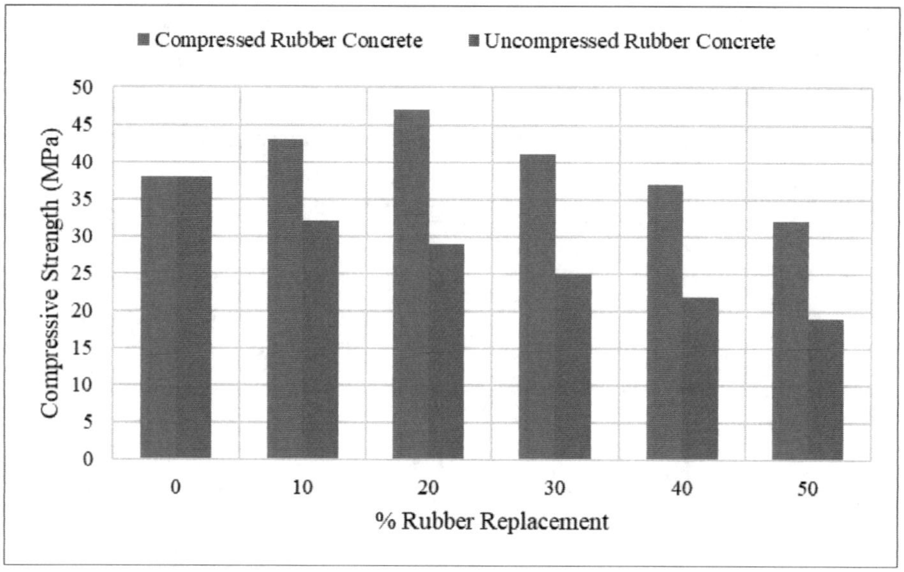

Figure 1. Compressive Strength vs % Rubber Replacement in Concrete Specimens

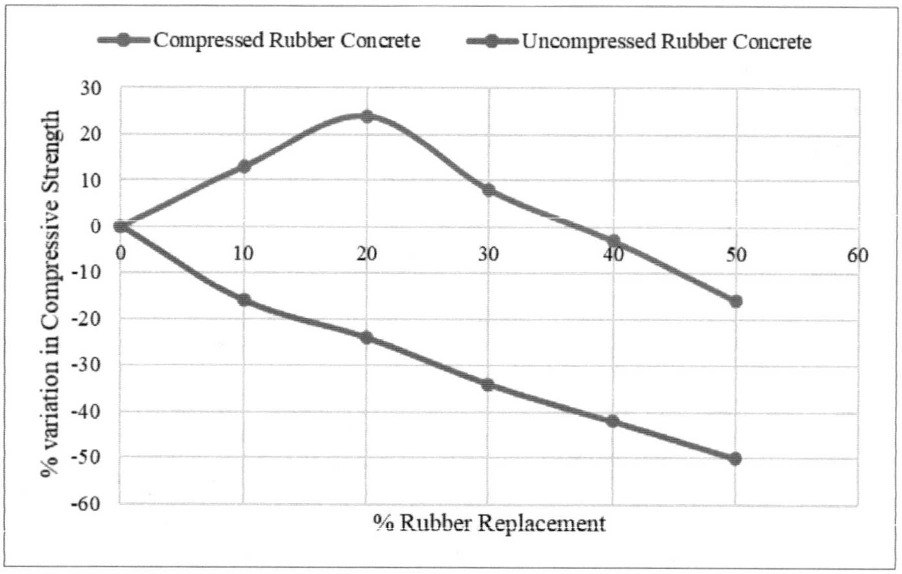

Figure 2. % variation in Compressive Strength vs % Rubber Replacement in Concrete Specimens

3.2. Split Tensile Strength

Figure 3 displays the effect of the replacement by the rubber waste on the split tensile strength of the concretes at 28 days. The reduction of the split tensile strength is noticeable in uncompressed rubber concrete specimens as the rubber replacement percentage is increased [17]. For instance, the UCR50 has split tensile strength reduced by 28% as compared to CC. A reduction in the split tensile strength of uncompressed rubber concrete is exhibited when the rubber content is increased so that a reduction by 56% at a rubber replacement of 50% takes place (Figure 4). For compressed rubber concrete, split tensile strength increased for a rubber replacement up to 20%. For further rubber replacement, split tensile strength has decreased in comparison with conventional concrete, irrespective of the compression technique used. Other researchers have also noticed a decrease in split tensile strength at higher rubber replacement dosage [18]. The reduction of tensile strength is caused by the same factors as affecting compressive strength that mainly includes an increase in porosity due to weak interaction between rubber and binder materials. The impact further intensifies with the increasing replacement dosage of rubber particles leading to the appearance of cracks on the application of load. The split tensile strength of CRC10, CRC20, and CRC30 increase by 13%, 22%, and 6% in comparison to CC. Since the maximum increase in split tensile strength is attained for CRC20, the optimum replacement percent of rubber is 20%. The split tensile strength of CRC40 is almost comparable to that of CC illustrating the reliability of the compression method to improve the rigidity of the rubber concrete.

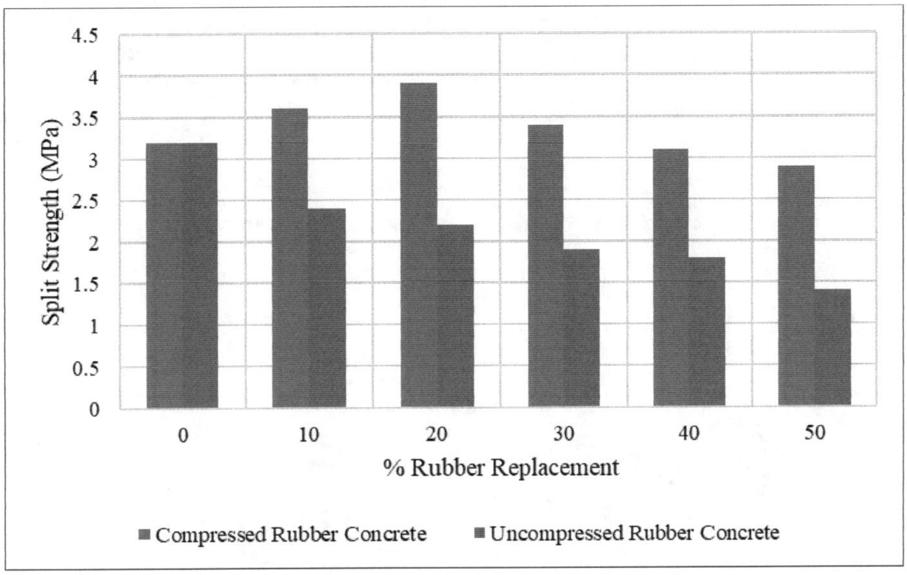

Figure 3. Split Tensile Strength vs % Rubber Replacement in Concrete Specimens

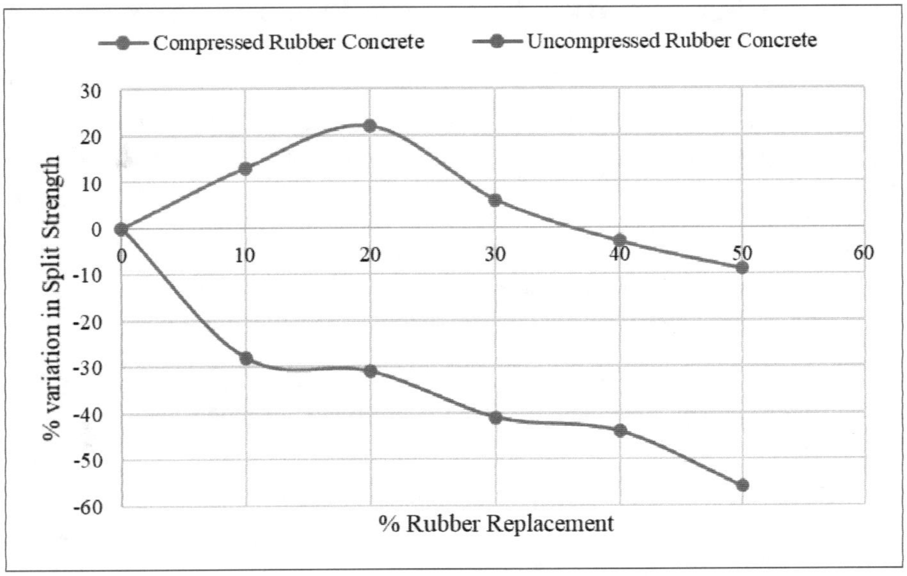

Figure 4. % variation in Split Tensile Strength vs % Rubber Replacement in Concrete Specimens

3.3. Abrasion Resistance

Figure 5 illustrates the impact of rubber replacement on the abrasion resistance of concrete in terms of depth of wear at 28 days. A decrease in depth of wear was observed for the concrete as the percentage of rubber increases irrespective of the compression treatment. This observation can be attributed to the flexible nature of rubber particles that acts as a shield for wear due to abrasion [16]. The depth of wear for CC was observed as 1.1 mm while at the 30% replacement level of rubber, a decrease of 59%, and 31% was observed in the case of CRC30 and URC30 respectively. Further, the abrasion resistance was found to improve with increasing rubber replacement level from 0-50% [19]. Although, the auxiliary chipped rubber waste improved the abrasion resistance of uncompressed rubber concrete as compared to conventional concrete, yet the abrasion resistance of compressed rubber concrete was better as compared to uncompressed rubber concrete for the same level of rubber replacement (Figure 6). For instance, in comparison to CC, a decrease of 13% was observed for URC10, while a decrease of 41% for CRC10. In case of compressed rubber concrete, abrasion resistance of CRC10 was similar to that of URC50 indicating the positive impact of compression technique. Thus, the results signify the role of compression technique in enhancement of mechanical properties of rubber-based concrete. Further, the results are within permissible limits confirming to the suitability for use of chipped rubber waste in concrete [20].

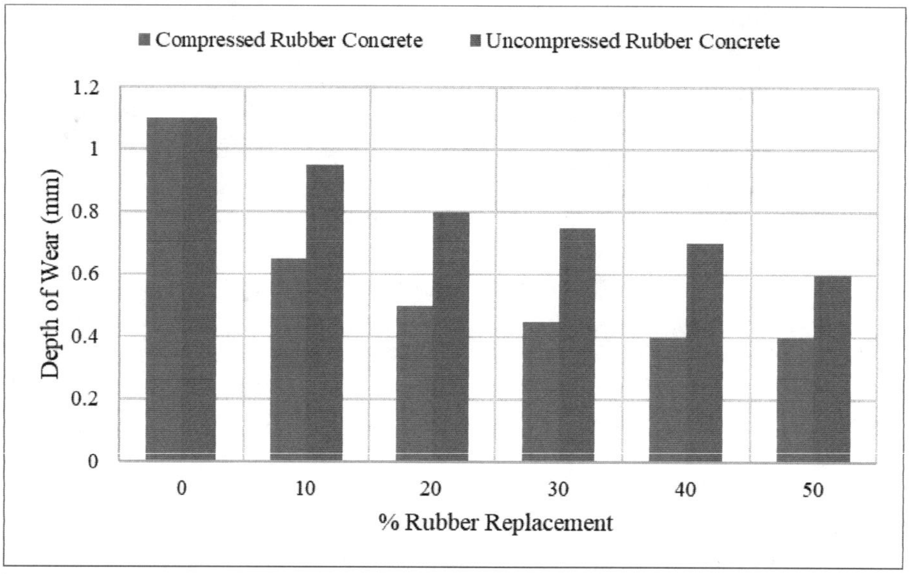

Figure 5. Depth of Wear vs % Rubber Replacement in Concrete Specimens

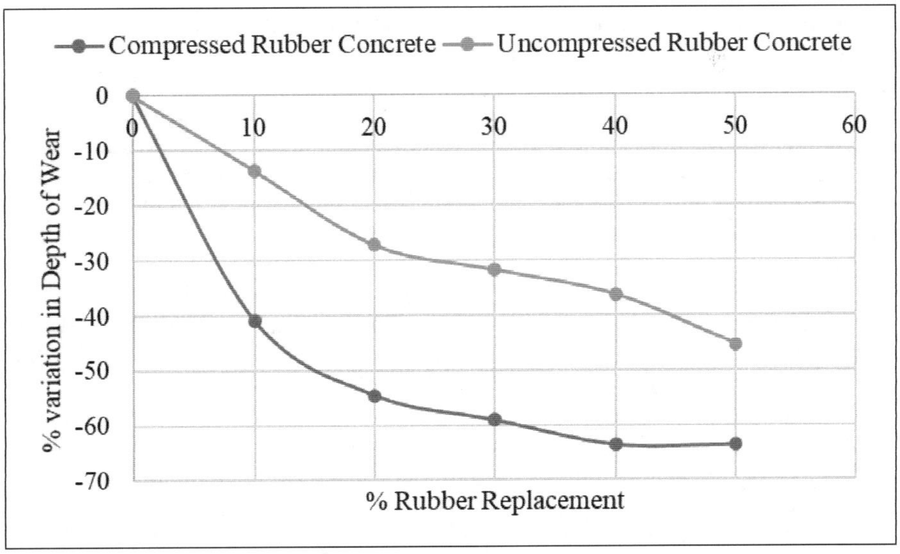

Figure 6. % variation in Depth of Wear vs % Rubber Replacement in Concrete Specimens

3.4. SEM Analysis

The microstructures of conventional concrete CC, uncompressed rubber concrete URC10, and compressed rubber concrete CRC20 has been shown in Figure 7. In the case of compressed rubber concrete, fewer pores and microcracks are observed along with denser microstructure than CC and URC10. The denser microstructure of the compressed rubber concrete with reduced pores and microcracks results in an enhanced strength instigated by the compression pressure used during casting to the concrete. Other researchers have also related the pore structure and strength of concrete [12]. Due to poor binding interactions between rubber and binder components in the matrix, the interfacial transition zone is weak in the case of uncompressed rubber concrete [20]. The pre-compression of concrete leads not only to the void condensation but also increases the rigidity of materials in concrete. The weak interfacial transition zone suffices the origin of microcracks in the cement matrix. As the interaction between the binder and the other constituents in the case of compressed rubber concrete gets improved due to the reduced pores, it leads to a denser microstructure of the compressed concrete as compared to the uncompressed type of concrete [18].

Figure 7. Micrographs of (a) CC; (b) URC10; (c) CRC20 specimens

4. Conclusion

This study was mainly done to signify the usage of the waste chipped rubber in the concrete industry. For achieving this goal, the compression technique was used to increase the compactness of concrete.

Waste chipped rubber was used as a replacement of coarse aggregates at a replacement level from 0% to 50%. Depending upon the research work some conclusions and suggestions are drawn:

1. The compressive strength test as well as the split tensile strength test of the compressed rubber concrete gives excellent results as compared to conventional concrete up to a replacement of 30%. The maximum result was found at a lower replacement percentage that is at 20% replacement. This type of concrete gives satisfactory results up to 30 % replacement but after this, the bond between the binder and the waste particles weakens which leads to the decrease in the strength so the recommendation from this research work is to use compressed rubber concrete up to a replacement of 30%.
2. Rubber replacement was found to improve abrasion resistance. The microstructure of the compressed concrete is far denser as compared to the conventional concrete and uncompressed rubber concrete. Compression casting densifies the concrete and leads to the condensation of the concrete.
3. The compressive strength and the split tensile strength of the uncompressed concrete were found to be poor as compared to compressed concrete and even with reference to the natural aggregate concrete. So it is advisable not to implement this type of concrete as its life as well as the strength aspects lead to the failure of the whole structure.
4. The compression technique should be implemented in the case of precast members where strength as well as the life of the structure is a major concern.

References

[1] Sofi A 2018 Effect of waste tyre rubber on mechanical and durability properties of concrete – A review *Ain Shams Eng. J.* **9** 2691–700

[2] Aiello M A and Leuzzi F 2010 Waste tyre rubberized concrete: Properties at fresh and hardened state *Waste Manag.* **30** 1696–704

[3] Thomas B S, Gupta R C and Panicker V J 2016 Recycling of waste tire rubber as aggregate in concrete: durability-related performance *J. Clean. Prod.* **112** 504–13

[4] Bravo M and de Brito J 2012 Concrete made with used tyre aggregate: durability-related performance *J. Clean. Prod.* **25** 42–50

[5] Senin M S, Shahidan S, Leman A S, Othman N, Shamsuddin S M, Ibrahim M H W and Mohd Zuki S S 2017 The durability of concrete containing recycled tyres as a partial replacement of fine aggregate *IOP Conf. Ser. Mater. Sci. Eng.* **271**

[6] Turatsinze A and Garros M 2008 On the modulus of elasticity and strain capacity of Self-Compacting Concrete incorporating rubber aggregates *Resour. Conserv. Recycl.* **52** 1209–15

[7] Huang W, Huang X, Xing Q and Zhou Z 2020 Strength reduction factor of crumb rubber as fine aggregate replacement in concrete *J. Build. Eng.* 101346

[8] AbdelAleem B H and Hassan A A A 2018 Development of self-consolidating rubberized concrete incorporating silica fume *Constr. Build. Mater.* **161** 389–97

[9] Roychand R, Gravina R J, Zhuge Y, Ma X, Youssf O and Mills J E 2020 A comprehensive review on the mechanical properties of waste tire rubber concrete *Constr. Build. Mater.* **237** 117651

[10] Thomas B S, Gupta R C, Mehra P and Kumar S 2015 Performance of high strength rubberized concrete in aggressive environment *Constr. Build. Mater.* **83** 320–6

[11] Arteaga-Arcos J C, Chimal-Valencia O A, Yee-Madeira H T and D??az De La Torre S 2013 The usage of ultra-fine cement as an admixture to increase the compressive strength of Portland cement mortars *Constr. Build. Mater.* **42** 152–60

[12] Rashad A M 2016 A comprehensive overview about recycling rubber as fine aggregate replacement in traditional cementitious materials *Int. J. Sustain. Built Environ.* **5** 46–82

[13] Onuaguluchi O and Panesar D K 2014 Hardened properties of concrete mixtures containing pre-coated crumb rubber and silica fume *J. Clean. Prod.* **82** 125–31

[14] Copetti C M, Borges P M, Squiavon J Z, da Silva S R and de Oliveira Andrade J J 2020

Evaluation of tire rubber surface pre-treatment and silica fume on physical-mechanical behavior and microstructural properties of concrete *J. Clean. Prod.* **256** 120670
[15] Eisa A S, Elshazli M T and Nawar M T 2020 Experimental investigation on the effect of using crumb rubber and steel fibers on the structural behavior of reinforced concrete beams *Constr. Build. Mater.* **252** 119078
[16] Choudhary S, Chaudhary S, Jain A and Gupta R 2020 Valorization of waste rubber tyre fiber in functionally graded concrete *Mater. Today Proc.* 3–8
[17] Ganjian E, Khorami M and Maghsoudi A A 2009 Scrap-tyre-rubber replacement for aggregate and filler in concrete *Constr. Build. Mater.* **23** 1828–36
[18] Wu Y, Kazmi S M S, Munir M J, Zhou Y and Xing F 2020 Effect of compression casting method on the compressive strength, elastic modulus and microstructure of rubber concrete *J. Clean. Prod.* **264** 121746
[19] Gesoılu M, Güneyisi E, Khoshnaw G and Ipek S 2014 Abrasion and freezing-thawing resistance of pervious concretes containing waste rubbers *Constr. Build. Mater.* **73** 19–24
[20] Gupta T, Chaudhary S and Sharma R K 2014 Assessment of mechanical and durability properties of concrete containing waste rubber tire as fine aggregate *Constr. Build. Mater.* **73** 562–74

IOP Conf. Series: Materials Science and Engineering **961** (2020) 012003 doi:10.1088/1757-899X/961/1/012003

Strength Behaviour of Clay Soil Stabilized With Lime

M Arun kumar[1,a] , P Kulanthaivel[2] , S Selvapraveen[3,b], S Vinodhkumar[4] , V Naveenraj[5]

[1,2,4,5]Assistant Professor, Department of Civil Engineering, Kongu Engineering College,Erode,India
[3]PG Student, Department of Civil Engineering, Kongu Engineering College, Erode 638060, India.

[a)] rsmanoharun@gmail.com
[b)] ksselvapraveen@gmail.com

Abstract. Geotechnical Engineers are searching for new sustainable material, which can be suitable and effectively used to improves the engineering properties of clay soil. In this research lime is used as stabilizing material in the variable percentage of 1%,3%,5%,7%,9%,10%,12% and 15%. Lime is used as the admixture for soil stabilisation. Swelling pressure of the soil mixed with various proportions of lime such as 1, 7, 12 and 15%. Unconfined compression test, Plastic limit, Liquid limit, standard compaction proctor test, free swell index test and one-dimensional consolidation test are conducted to evaluate the clay soil strength in the laboratory. The experimental investigation results show that maximum strength improvement was achieved in the 12% lime mixed with clay soil and the strength improvement in the range of 300kPa. The results of the swelling test showed that applying lime to the soil decreases the swelling nature of clay soils.

Keywords: Clay soil, Lime, Unconfined compression test, Free swell index test

1. Introduction

Clay soil undergoes swelling when water is getting added to themwhen they are dry. On drying, all clays shrink until the limit of shrinkage is reached. There are also special clays in which this problem is very pronounced, called expansive clays. The two main factors to consider in these swelling soils are firstly the amount of swelling and secondly the swelling pressure during swelling. It depends on the composition of the mineralogical and the amount of clay particles found in the soil. It is well recognized that because of the structural differences, montmorillonite absorbs more water than kaolinite. It is also

IOP Conf. Series: Materials Science and Engineering **961** (2020) 012003 doi:10.1088/1757-899X/961/1/012003

established that montmorillonite can be formed from limestone, while in the presence of magnesium ions and lack of leaching the black cotton soil of India has been formed from igneous rocks such as basalt in an alkaline environment. These soils exist in some parts of southern India, Bhopal and some other places [1-9].

Soil that exhibits a high volume change when it comes to water contact, called expansive soil. Black cotton soil is kind of expansive soil. This soil shows high plasticity, high free swell, high unconfined compression strength, high cohesion and lower penetration resistance against the CBR plunger. The liquidity property of the soil is high[10].

This soil forms cracks without prior warning. This cracks has average 7mm wide and 300mm deep. In south India structure of the black cotton, the soil is like a lentil seed. This type of structure is usually found near the Hyderabad. The subsurface shape of this soil is like a double-convex lens. At the start of the wet season, the infiltration rate is high in black cotton but the water content increases the infiltration rate drops in black cotton soil. Jewitt et al. investigated some African soils and found infiltration rates ranging from 0.2 mm / h ~ 0.5 mm / day, soil cracks are fixed and the profile becomes completely wet. The surface runoff occurs in such extremely low infiltration rate. In dry condition, Black cotton soil reveals high bulk density and low value in swollen stages. The bulk density of the soils may be (1~2) gm/cm3, depending on the dampness condition. (Jewitt et al.). American origin Black cotton soil having a bulk density range from (1.81~2.08) gm/cm3. Bulk density of Huston clay range from (1.59~2.1) gm/cm3. Black cotton soil in India (Hyderabad) expanded up to 60% when gets soaked from a dry state. The clay has a special property expect all the soils that do not have is known as plasticity. Water is electrical polar molecule, the polar molecule is a molecule in which centre of gravity of positive charges (Hydrogen) and centre of gravity of negative charge (Oxygen) which does not coincide. This causes water to the charged compound and made them react with the clay minerals. This cause the occurrence of an electrical double layer in the clay, this is the reason for the clay shows the special property named plasticity. Montmorillonite is the major component for the black cotton soil to show the high swelling (increase in volume) and shrinking (decrease in volume). Cracking occurs in the soil due to the shrinking of the soil. Shrinking is more dangerous than swelling in the structural point of view. Swelling cause the major problems in the lighter structures but shrinkage cause all types of structures [1-5].

Black cotton soil pH is generally varied from the 7.5 to 8.5. The pH of the black cotton soil indicated that the soil is alkaline. The alkalinity of the soil is because of the presence of the Calcium, Caco3 and magnesium salts in the soil. In Tropical countries like India, more fertilizers are used in the field, this will result in an increase of the sodium salts content in the soil. This will destroy the montmorillonite structure in the soil and pH will be increased to 9.5.

2. Materials and Methods

2.1 Soil Sample

Soil samples are obtained in the Coimbatore region at a depth of 0.5 m from the farm field. Excavation depth indicates that the chemicals and fertilizers do not affect at that depth. Soil is collected and dried out for experimental purposes. The basic soil properties were calculated using Indian Standard soil testing procedures and the results of the tests were tabled in Table 1. On sieve analysis over 50% of soil passes on the 75-micron sieve it shows that soil is either silt or clay. The plasticity index of the black cotton soil is 37. As per the A-Line graph, Plasticity is obtained as 36.5. Actual Plasticity Index is greater than A-Line Plasticity index so it confirms soil is clayey soil. Actual Liquid limit of the clayey soil is 70 % so that the soil is high compressible clay soil and it is indicated as CH. Free swell index of the soil is about 65 % so it confirms that the soil is expansive soil [11-14].

Table 1. The basic properties of soil sample

S.No	Property	Test Results
1	Specific gravity	2.69
2.	Particle size less than 0.075 mm	70%
3.	Liquid Limit	60%
4.	Plastic Limit	31%
5.	Plasticity Index	35%
6.	Soil Classification	CH
7.	Maximum Dry Density	1.32 g/cc
8.	Optimum Moisture Content	15 %
9.	Unconfined Compression Strength	35.46 kN/m^2
10.	Free Swell Index	63 %

2.2 Lime

Lime is an inorganic mineral containing calcium consisting mainly of oxides, and hydroxide, typically calcium oxide and calcium hydroxide. Such materials are also used as construction and engineering materials in large quantities (including cement, concrete, limestone products and mortar). Lime is commercially available in the market and in this study, the lime was purchased from VR chemicals, Erode. Lime properties were presented in Table 2.

Table 2. The basic properties of lime

S.No	Property	Results
1.	Specific gravity	2.4
2.	CaO	72.11
3.	$Al_2O_3 + SiO_2$	25.75
4.	MgO	3.19
5.	Loss of Ignition	2.45
6.	Bulk density	2.0 g/cc

3. Sample Preparation

3.1 Unconfined Compressive Strength

The UCC is a special triaxial test type in which the containing pressure is zero. Only clayey soil that is able to stand without any containment can be tested.

The sample shall be prepared for the test before the test is to be conducted. First SPCT is done and optimum soil moisture content is determined. The new soil sample is taken and combined with the soil's OMC and compacted as per SPCT. Then the SPCT mould is kept in the sample extractor in which the sample of the soil to be extracted for the UCC test. It is noted that the height of the sample used for the UCC at least equal to the twice the diameter of that sample.

The sample is to be placed in the fixed bottom plate in the UCC apparatus. Proving is used to measure the compressive force applied in the specimen. The specimen has two plates that have one seated. The specimens are placed on the base plate to be in contact with the upper plate. The dial indicator and proving ring are set at zero. As the motor starts to rotate, the rotating upper plate applies compressive force. The strain is measured in dial gauge and the upper plate rotates in such a manner that the strain in the specimen does not exceed 2 percent per minute. The shearing continues until the specimen fails or until 20% of the axial strain occurs, whichever is earlier.

In UCC, soil cohesion is equivalent to half of an unconfined strength of the compression. The failure plane in the soil makes the 45 degree to the horizontal this is because a flexible material applied pure compressive load it fails at 45 degree. If the failure plane is drawn for the Mohr circle it shows semicircle and the minor principal plane stress is zero. In Mohr circle, the failure plane angle is 90 degree. It is again proof of that Mohr circle failure plane angle is half that in the material.

Stress-Strain graph is drawn for the Mohr circle. Stain in X-axis and Stress in Y-axis. The materials Unconfined compressive strength has been taken from the stress-strain graph of the UCC.

3.2 Consolidation Test

Consolidation test is used to evaluate soil consolidation rate and magnitude when the soil is laterally constrained and axially loaded. Often referred to as One-Dimensional Compression Test or Standard Oedometer is the consolidation test. This test is done on specimens of saturated soil, particularly in cohesive soils. Saturate two porous blocks, submerging them into distilled water for about 4 - 8 hours. Wipeout excess space. The consolidometer fittings to be mounted shall be moistened.

Arrange the consolidometer on top and bottom of the instrument with the soil specimen and porous stones, providing a filter paper between the soil and the porous stone test. Locate a pressure pad on a porous stone centred on top. Place the mould onto the load frame and position it so that the applied load is axial. Insert the dial gauge for vertical compression test of the specimen. The dial gauge holder should be calibrated to allow the dial gauge to run in the begging of its releases, leaving ample room, if any, for soil swelling.

Attach the mould assembly to the water reservoir, and allow saturation of the sample. The level of water in the reservoir will be approximately equal to that of the soil specimen. Initial load was applied to the assembly. This load should be magnitude not less than 50g/cm3. The loading must be allowed to stand until the dial gauge readings adjust for two consecutive hours or a maximum of 24 hours.

Notice the reading of the final dial under the p load. Apply 1st intensity load 0.1kg/cm2 and start the stopwatch simultaneously. Record the results of dial gages at different intervals of time. The readings of the dial gauge are taken before consolidation hits 90 per cent. The primary consolidation is achieved gradually in 24 hours.

This test can also be used to find the swell pressure of the expansive soil. Swell pressure is the pressure by which expansive soil attains the original volume. Swell pressure is found by plotting a graph between swell pressure (X-axis) and the change in volume (Y-axis). The test is repeated for a various percentage of lime.

4. Results and Discussion

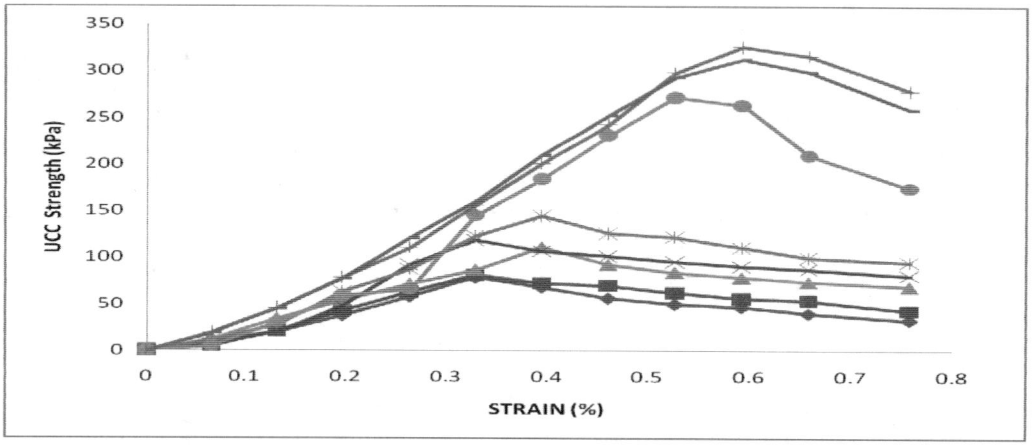

FIGURE 1. Different dosage of lime on clay soil (1 day curing)

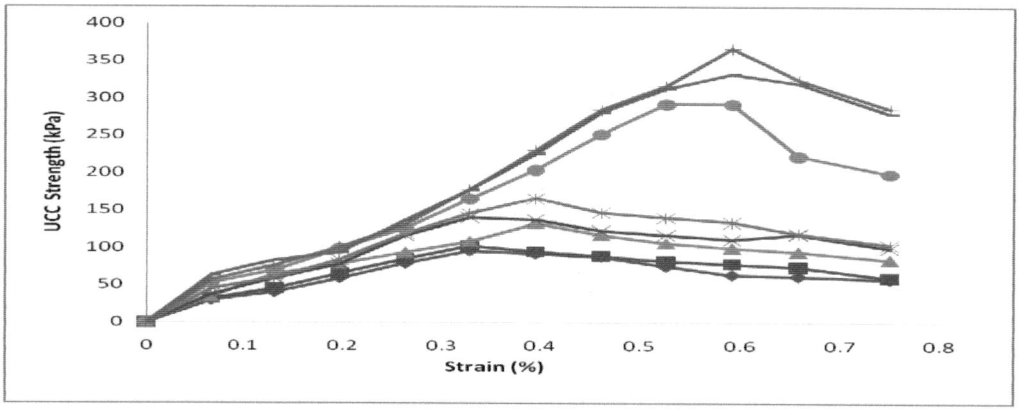

FIGURE 2. Different dosage of lime on clay soil (3 days curing)

Fig.1 and Fig.2 show that the effect of lime content on clay soil with one day and three days curing. The maximum strength of the soil is achieved in one-day curing was 327 kPa and for three days curing maximum strength was 367 kPa. The strength improvement was in the range of 10 times greater than the untreated soil for one day curing and for three days curing the strength improvement was 11.35 times the untreated soil. The UCC strength of the clay soil improved with increasing the lime content. But at a certain percentage of lime (e.g. 15%) the UCC strength decreases. Additional application of lime also reduces soil strength. The test results show that greater strength has been achieved at 12% lime for both one and three days curing period. The lime content added to the wet soil increases the pH of the soil which is responsible for strength improvement. The pH of lime mixed soil having pH in the range of 11 to 12. The test results showed that as the lime content applied to the soil raising the liquid and plastic limit to 12% of the lime and above that 12% of the lime applied, all limits started to decrease. The increase in Atterberg 's limits can be attributed to soil flocculation after lime content is applied. In this process, the clay particles appear to agglomerate and form huge aggregates. However, increasing the lime level by more than 12% results in a reduction of the liquid limit with an increase in plasticity for the samples being treated. The reason for such behaviour can be discussed via cation exchange capacity, where calcium ions are released from the addition of lime which tends to replace other cations around the surface of the clay particle resulting in a reduction of double-layer thickness.

Kulanthaivel et.al (2020) demonstrated that the maximum strength of intermediate compressible clay soil stabilized with nano-silica was 294.461kpa. In the current study, the maximum unconfined compressive strength of high compressible clay soil treated with lime content was 367kpa. When comparing the above two studies indifferent results were identified because of the type of admixture, type of soil, treating method used in two studies were different[15].

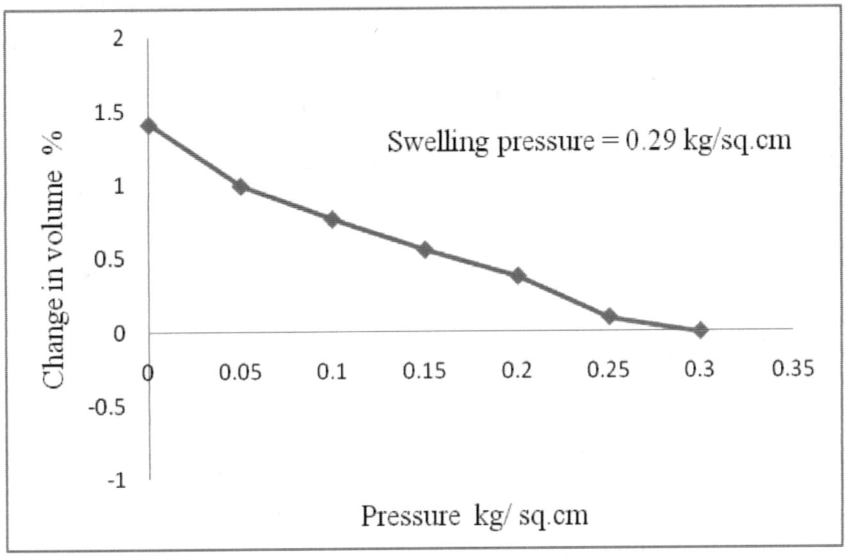

FIGURE 3. Consolidation test for untreated soil

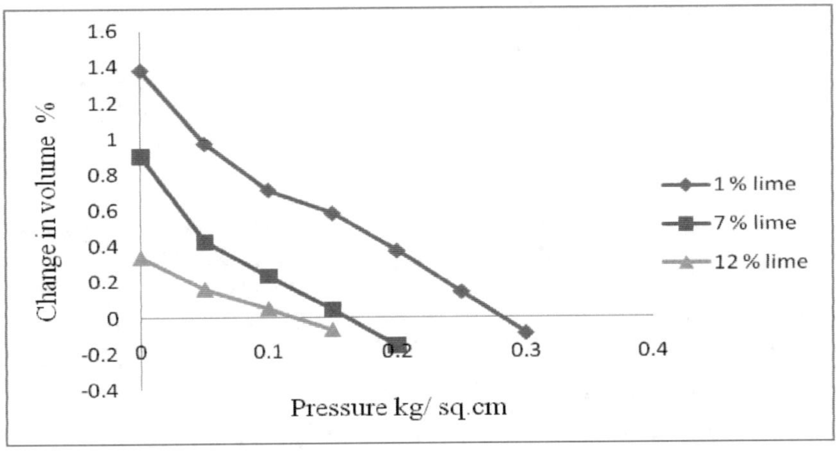

FIGURE 4. Consolidation curves for 1%, 7% and 12% lime

One dimensional consolidation test can be used to find out the pressure at which swelling is arrested. The test is conducted for both untreated and treated samples of (1%, 7%, and 12% lime). The increased volume of untreated soil was in the range of 1.6. Adding lime content 1%, 7% and 12% decreases the volume of the soil 1.4, 1.0 and 0.4 respectively. The test results show that pressure required to arrest the swelling for the untreated sample is high when compared with the treated samples.

From the Fig.3 and Fig.4, the test data demonstrated that increasing the dosage of lime content reduces the consolidation pressure. At 1% lime content the pressure was 0.3kg/m^2 and for 7% lime content the consolidation pressure was 0.2. The lime content increased from 7% to 12% the consolidation pressure was reduced from 0.2 to 0.1 kg/m^2. Similarly, the volume was reduced from 1.6 to 0.4 when the lime content increased from 1% to 12%.

5. Conclusion

The clay soil stabilized with lime content gives the following conclusions.

(i) The maximum UCC strength was achieved by optimum lime content of 12%. Beyond this percentage addition of the lime, content decreases the strength behaviour of soils.

(ii) The maximum UCC strength was achieved as 327 kPa for one-day curing and 367 kPa for three days curing.

(iii) The swelling of soil was reduced at 12% lime content and the final volume was reduced in 1.6 to 0.4.

(iv) Addition of lime content to the clay soil increases soil strength characteristics and reduces swelling.

References

[1] Rama Vara Prasad, C., et al., *Swelling characteristics of soils subjected to acid contamination.* Soils and Foundations, 2018. **58**(1): p. 110-121.

[2] Emarah, D.A. and S.A. Seleem, *Swelling soils treatment using lime and sea water for roads construction.* Alexandria engineering journal, 2018. **57**(4): p. 2357-2365.

[3] Stoltz, G., O. Cuisinier, and F. Masrouri, *Multi-scale analysis of the swelling and shrinkage of a lime-treated expansive clayey soil.* Applied Clay Science, 2012. **61**: p. 44-51.

[4] Onyelowe, K., et al., *Scheffe optimization of swelling, California bearing ratio, compressive strength, and durability potentials of quarry dust stabilized soft clay soil.* Materials Science for Energy Technologies, 2019. **2**(1): p. 67-77.

[5] Latifi, N., et al., *Micro-structural analysis of strength development in low-and high swelling clays stabilized with magnesium chloride solution—A green soil stabilizer.* Applied Clay Science, 2015. **118**: p. 195-206.

[6] Jamsawang, P., et al., *Laboratory investigations on the swelling behavior of composite expansive clays stabilized with shallow and deep clay-cement mixing methods.* Applied Clay Science, 2017. **148**: p. 83-94.

[7] Thakur, V.K. and D.N. Singh, *Rapid determination of swelling pressure of clay minerals.* Journal of Testing and Evaluation, 2005. **33**(4): p. 239-245.

[8] Chen, Y.-G., et al., *Combined thermal and saline effects on the swelling pressure of densely compacted GMZ bentonite.* Applied Clay Science, 2018. **166**: p. 318-326.

[9] Guney, Y., et al., *Impact of cyclic wetting–drying on swelling behavior of lime-stabilized soil.* Building and environment, 2007. **42**(2): p. 681-688.

[10] Langroudi, A.A. and S.S. Yasrobi, *A micro-mechanical approach to swelling behavior of unsaturated expansive clays under controlled drainage conditions.* Applied Clay Science, 2009. **45**(1-2): p. 8-19.

[11] 4, I.-P., *Indian Standard Methods of Test for Soils: Grain Size Analysis.* 1985, Bureau of Indian Standards New Delhi. p. 73-91.

[12] IS, *Methods of test for soils: Determination of liquid and plastic limit.* 1985, Bureau of Indian Standards New Delhi.

[13] BIS, *IS 2720 (Part 40)-1977: Determination of free swell index of soils.* 1977, Bureau of Indian Standards New Delhi, India.

[14] IS and I.-P. 3, *Methods of test for soils–determination of specific gravity.* 1980, Bureau of Indian Standards New Delhi, India.

[15] Kulanthaivel, P., et al., *Experimental investigation on stabilization of clay soil using nano-materials and white cement.* Materials Today: Proceedings, 2020.

IOP Conf. Series: Materials Science and Engineering **961** (2020) 012004 doi:10.1088/1757-899X/961/1/012004

Mechanical Properties of Hybrid Polymers for Break pad Applications.

Ravi Raj.V[1,a], Vijaya Ramnath.B[2],Srinivsan.R*[3],Sethuvelappan.P[4],Ramanan[5]

[1]Research Scholar, Department of Mechanical Engineering, Anna University, Chennai
[a]Associate Professor, Department of Mechanical Engineering, Sri Sairam Engineering College, Chennai
[2]Professor, Department of Mechanical Engineering, Sri Sairam Engineering College, Chennai
[3]Professor, Department of Mechanical Engineering, Sri Krishna College of Technology, Coimbatore
[4]UG Scholar, Department of Mechanical Engineering, St. Joseph's College of Engineering, OMR, Chennai
[5]Engineer, Synce Engineering Service, Chennai

*E-mail: vlbstini@gmail.com

Abstract. Non wear out brake emission is one of the significant issue concerns with health and the environment. In brakes performance, wear debris is released in the atmosphere and pollute the environment. The brake pad is classified into two categories mainly asbestos and non-asbestos. As per the report of WHO and Environment Protection Agencies, some of the metals oxidized and produces toxic gaseous. Toxicity leads and creates a bad effect on human health. This work focuses on remedies over environmental and health issues because of brake pad emission. It is important to reduce the percentage of toxic elements at source and develop novel material compositions that are less dangerous to the environment and human health. Also, it satisfies all requirements of a brake pad material as per international standards. From this investigation, we found an alternate material for brake bad applications. The natural fibers reinforcement material is zero toxic. So, the materials will not create any environmental and human health issues.

Key Words: Recycle Material, Hazard Material, Break Pad Material, Environmental Health issue

1. Introduction

Particulate matter (PM) in the atmosphere is an important component of air pollution that can cause adverse health effects PM comprises a complex mixture of components derived from different sources, with motor vehicle emissions being one of the most important sources in urban areas.[1] Several studies have shown that airborne brake wear particles often differ considerably from the bulk friction material.[2] The emissions and compositions of brake wear particles vary depending on several factors: (a) brake friction material parameters, for example, the contents of non-asbestos organics (NAOs), low-alloy steel, semi-metallic components, Glass composites, and rotor and drum parts[3],Brake assembly type, including discs, drums, assembly sizes, surface structures, and depth of grooves and vehicle operating conditions, including initial speed, deceleration, pressure, torque, and brake

temperature [4]. Studies into the mass balance measurement of airborne brake wear particles following braking events have estimated that 30–50% of generated brake wear debris becomes airborne [5].
It is also thought that brake wear substances can be generated from hydroGlasss derived from the gasification of resins in NAO friction materials [6]. Metallic components decompose under tribo-reduction, which is defined as the reduction of metal oxides associated with friction[7]. However, using actual vehicles and/or roadway tests may result in PM samples that are contaminated by re-entrained road dust and/or tire wear particles [8]. Further research is required to properly determine the emissions of brake wear particles, including the careful assessment of friction under transient driving cycles[9] .Brake wear particle mass emission studies are typically performed according to initial speed/deceleration driving conditions and brake temperatures[10].On low-velocity impact behavior, the number of the Glass and aramid layers and fiber orientation were reinforced on the fabric hybrid composites with a stratified filled epoxy matrix. The impact tests were done with the drop weight impact system at 90.629 J of energy level. The areas which underwent damage were analyzed by the visual inspection and tomography images. The results indicate that the matrix properties have a great influence over the fracture mode of the hybrid composites, while the fault degree of the damaged areas depending on the fiber orientation. The high impact resistance was obtained in the case of hybrid composites with 0° ply orientation [11]. The Glass fiber reinforced composites perform better in fatigue loading, in comparison to glass fiber reinforced composites; The effect of polymer aging was also evaluated through thermal aging of neat resin specimens.[12]. The flexural performance, under both static and fatigue loading, of a glass fiber/Glass fiber hybrid polymer matrix composite material was studied. It indicated that exposure to 180 °C in the air for [13] months resulted in improved fatigue performance of the composite, with mildly diminished static flexure strength, after six months the performance resulted in moderately degraded fatigue performance and a further reduction in flexure strength, while severely reduced performance (both static and fatigue) for 12 months aging time was found. The mild improvement in fatigue performance for 3-month aged specimens attributed to the post-cure of the composite. They have the potential to compete with traditional fiber in all thermal and mechanical properties. The thermal degradation of vetiver fiber composite had more thermal decomposition if they undergo acetylation and silane chemical treatment and could treatment and could have still modified the surface of the fiber. [14].In this work, Jute, Pineapple leaf fiber, and Glass fiber reinforced hybrid composites were prepared. The tensile and flexural properties of the composites with these fibers were found to be increased with fiber content, conforming to the reinforcing action of the fibers. Thus the composites of Jute, Pineapple leaf fiber, and Glass fiber epoxy composites were found to be light in weight, possessed better mechanical properties. Hence these composite materials can be used for applications such as automobile parts, electronic packages, building construction.[15].This paper characterized the impact response of a new hybrid composite material made using Type A (woven Glass fibers / UD flax fibers) and Type B (woven Glass fibers / cross-ply flax fibers) configurations in an epoxy matrix. The Type B composite had better overall impact performance as indicated by lower absorbed energy, higher penetration energy, smaller crack lengths, smaller indentation depths, smaller damage areas, lower temperature rise, and higher impact strength vs. the Type A version of the material. Both the Type A and B hybrid composite improved various impact properties compared to neat flax fiber composites reported in prior literature. The impact properties of the current composite may potentially have various research and industrial uses depending on the particular needs of the application.[16]. In a new hybrid polymer composite with banana, a natural fiber, as reinforcement and also to study the machining characteristics of the composite. Response surface methodology was used to analyze the influence of machining process parameters on the material removal rate and surface roughness. The effect of machining parameters on the material removal rate and surface roughness were obtained through regression equations and different plots obtained from ANOVA analysis, employed through MINITAB and ANFIS. It is concluded that feed rate has maximum influence on the material removal rate and surface roughness followed by speed and depth of cut. It is observed that cardanol and Cashew nutshell liquid-based composites have similar material removal rates but they vary significantly in surface roughness. Further scope of this investigation is to find the optimum values of feed rate, speed, and depth of cut

for material removal rate and surface roughness. It is also planned to investigate the various mechanical properties like tensile behavior, flexural behavior, dynamic behavior, etc. to suggest suitable applications.[17] In recent days, technology needs a high demand for the growth of future generations. Hence, to produce a good product which needs to satisfy the customers, it should be developed in an economical way, high quality with minimum possible time challenging among the competitors. Natural fiber-based polymer composites are the most shows the potential option for synthetic fiber and powerful polymer composites. It enhances the awareness of eco-friendly material makes it crucial to utilize natural fiber as potential reinforcement in polymer composites. Natural fibers are plentifully existing, biodegradable, and recyclable, which makes them conventional in the automobiles industries. The major negative aspect of testing is stretched time, the manufacturing cost is high, mistakenness, machine error, today, most bio composite researchers are moving toward computational methods to model the NFRPC to simulate the mechanical and thermal properties. Studied the tensile properties of flax/basalt hybrid composites with loading conditions.[21] They reported that the tensile properties of composites decreased with increasing aging conditions.[20] The rigidity and modulus diminished because of the embrittlement of the matrix materials. Comparative attributes were observed for sisal fiber reinforced PP composites [[18,19]] The affiliation between the cross-section of the natural fiber and the tensile properties. They modeled and evaluated the tensile performance of hemp fiber with dissimilar elliptical cross-sections. The results appeared that the tensile properties have a physically powerful manipulate on the degree of elasticity. The micro fibril angle and the viscoelastic properties played a very important role in the geometry of the natural fiber [22]. Presently, we have used natural fiber material as used in a wind turbine material. Because It is used less weight and observed high wind turbine energy efficiency.

2. Material and Method

Vetiver fiber belongs to the hibiscus family which is 4000 years old crop and originated in ancient Africa. Vetiver plant can grow with an average height of 1.5m to 4.5m tall in 5 months yielding 500-1000 kg of dry fiber/acre. Vetiver is very much suitable as biological resources, fossil fuels, wood pulp, etc due to its extensive application in the engineering field. GFRP consist of minute particles of fiberglass. The main advantage of using this fiber is due to its lightweight, robust, and strong. It can be easily molded into any shape. the particles are called filaments that are formed and bundled together to form a roving like structure. Bulk strength and weight properties are also very favorable when compared to metals. In the hand layup method, resins are impregnated into fabrics by manual feed which is mainly of roving form. The fibers and resins have impinged on the surface with the help of the rollers which has been fed into the mold. Then the fibers are made free under standard atmospheric conditions. This method is very simple and easy but it requires high skill to fabricate the composite. The fibers used are twisted GFRP& Vetiver fiber. in addition to that glass-reinforced polymer fiber. Glass fibers are laid at the top and bottom most of the composite laminate for better finishing purposes. Three different categories with three samples have been fabricated using hand layup method. All the fibers are laid in a normal direction and kept in a dry condition before it is fed to the laminate. Initially, the releasing agent is applied over the surface to remove the laminate easily. After applied, a thin layer of resin is applied ad then the glass fiber is laid on the surface. A weight of 5 kg is placed over the glass fiber to remove air bubbles if any and it is kept undisturbed for about 3 hours. Fibers are also dried in the normal condition to make the fibers moisture-free.

The below percentage ratio added vetiver and fibers

Table 1: Material Description

orientation	*Fibers*		*Matrix*
type 2	*Glass 90%*	*Vetiver10*	*Epoxy*

type 3	*Glass 80%*	*Vetiver20*	*Epoxy*

3. Manufacturing method

In the method of hand layup, the resins are impregnated into fabrics by the process of manual feed which is mainly of roving form. The rollers help the fibers and resins to get impinged onto the surface which has been fed into the mould. Then the fibers are made free under standard atmospheric conditions. It is one of the simpler and easiest methods but it requires high skill to fabricate the composite. The fibers used here are the glass and vetiver. For better finishing purposes the glass fibers are placed on both the top and bottom side of the composite. Three different categories with each containing two samples are fabricated using the hand layup method. All the fibers are laid in the normal direction and are kept in a dry condition before it is fed into the laminate. To remove the laminate easily the releasing agent is applied over it. A weight of 5 kg is placed over the Glass fiber to remove the air bubbles if any and is kept undisturbed for 3 hours. Also, fibers are dried in the normal condition to make the fibers moisture-free.The laminate thus obtained after removal are used in a variety of engineering applications.

4. Result and discussion

4.1. Tensile Test

For determining the tensile strength of the hybrid composite, a universal testing machine is used. The table shows that the tensile properties of the hybrid composite sample. The graphs have been plotted between the stress and the strain and also a comparison between categories concerning each property has been shown in Figure 1.

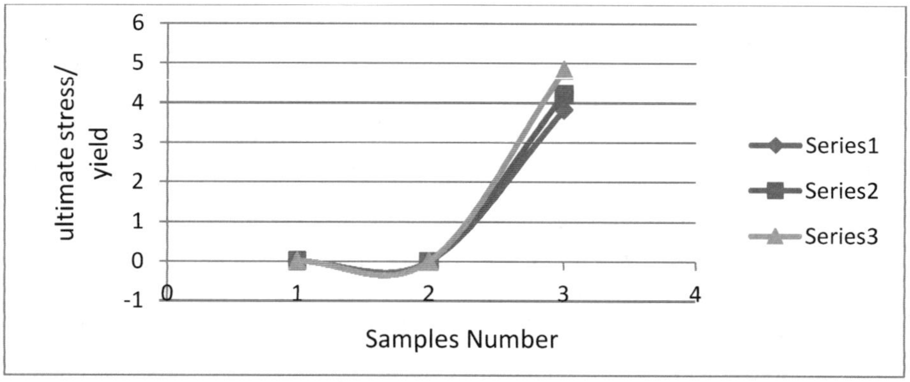

Figure 1: Tensile Test result

The above graph indicates the tensile test results. The results indicate that it has maximum ultimate stress of 4.9 Kn/m2 and the yield stress of 0.4 N. The graphs are shown in ductile fracture, less porosity occurs here hence the ultimate stress value increases. The maximum of the ultimate tensile value occurs at 5.0 Mpa. As the ultimate stress value increases the corresponding yield point also increases. The above graph shown is the correlated value of UTS and Yield stress values.

4.2. Flexural stress

The flexural test can be determined by a three-point bending test. Table 5 furnishes the experimental results of various parameters and also comparison has been done among the properties as shown in Figure 5. The graphs plotted as shown in Figure 5for all the categories are summarized. It is clearly understood that category 3 of sample 2 shows higher flexural strength with a maximum load of 1.65kN. The bending due to the flexural test has been restricted to the minimum level due to the presence of fibers that resist bending during loading conditions. Flexural modulus of 403.95MPa is recorded high when compared to category . The flexural result shown in figure2. The maximum displacement up to 1.4 mm. and the maximum stress 1.2 KN and maximum displacement up to 1 mm. Hence material are high elongation are to occur.

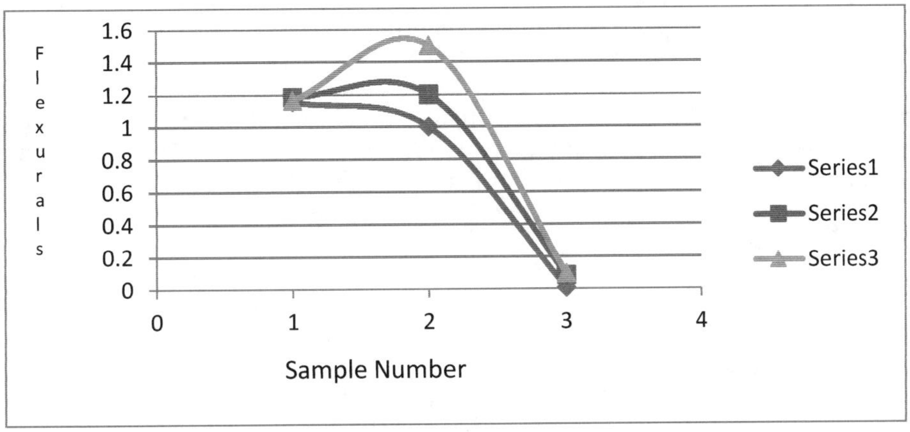

Fig 2: Flexural Test Result

As per the ASTM D368 standards, the flexural tests are undertaken, the flexural tool acting as axial stress. The axial stress varies proportionately to that of the increase in bending loads. The above figure indicates the bending stress, the maximum bending occurs at 1.5 kN, and having a maximum displacement up to 1 mm.

4.3. Double End shear

The experimental results obtained are tabulated below. Graphs between the load and the UTs are shown in Figure 6. The comparisons of various properties are also shown in Figure 6. It is found that superior property which has a break load of 4.04kN with a maximum displacement of 34mm due to the presence of glass and vetiver exists in the composite laminate which shows considerably good shear properties. Both the glass and vetiver fibers tend to improve the shear property proportionate with the matrix medium.

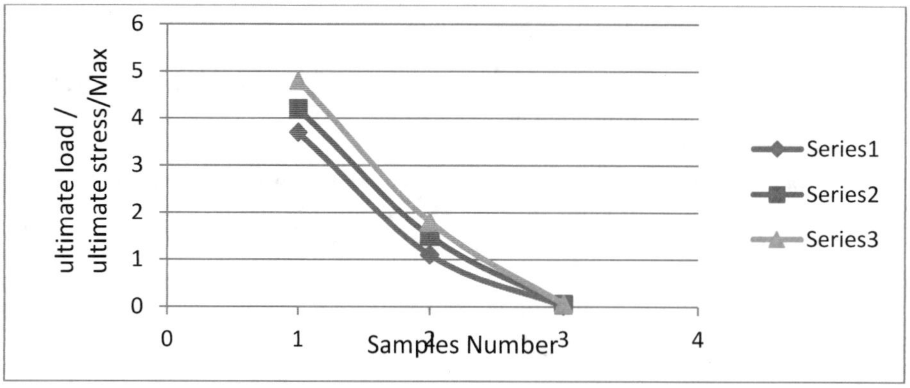

Fig 3: Double End shear Result

In this double end shear test, the test load acts at the axial points. The test results indicate that it has maximum ultimate stress of 4.2 N/mm and the maximum stress occurs at 4.9 kN. It has a maximum bending strength of 8 kN. The bending strength increases correspondingly to that of the increase in bending shear. This results in the material having higher durability.

4.4. Hardness

Hardness mainly depends on the material property in which the intended portion tends to absorb sudden shock which results in testing the hardness of the material. Category The hardness value is shown in a figure. The values are shown in up to 100 HRL. The material is shown in higher hardness. The material compares to metallic material are also equal.

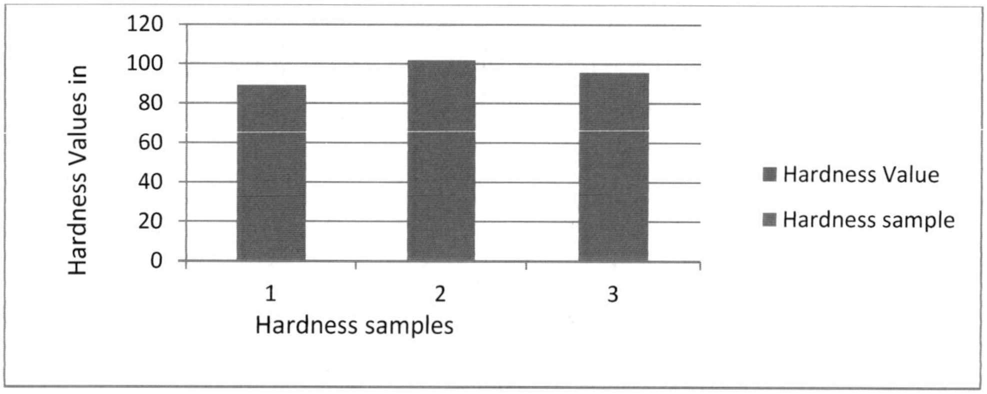

Fig 4 : Hardness Result

4.5.Wear Testing

This work tells us about the dry sliding wear behavior of Aluminium-Hexagonal Boron Nitride- Cubic Boron Nitride using a pin on disc apparatus (DUCOM™ make). Figure 4.1 shows the arrangement of the pin-on-disc apparatus. The disc material was made of EN-32 steel with a hardness of 65 HRC. The pin specimen is pressed against the disc at a specified load usually employing an arm and attached weights. The dry sliding wear tests were carried out at room temperature (30°C ± 3°C, RH 55 % ± 5%)

under the dry sliding condition following the ASTM G 99 standard. Cylindrical pins of 10 mm diameter and 25 mm long were machined and polished by the metallographic method. Immediately before testing, were cleaned and dried using acetone to remove all dirt and foreign matter from the specimens. The initial and final weight of the specimen was measured using a Mitutoyo™ to make an electronic weighing machine with an accuracy of 0.0001 grams. Wear measurement is carried out to determine the number of materials removed (or worn away) after a wear test, (and in reality after a part in service for some time). The material worn away can be expressed either as weight (mass) loss, volume loss, or linear dimension change depending on the purpose of the test, the type of wear, the geometry and size of the test specimens, and sometimes on the availability of a measurement facility. The specimen was fixed in the arm and the disc was made to rotate below the pin. Mass of the specimen before and after the test is measured using digital balance and mass loss is found out for each set of input parameters.

4.6.Wear Results

The Experiments were conducted according to the Taguchi's L27 Orthogonal Array- 3 level and 4 Factors. The Input Parameters and their Levels is given in Table 2. The result of the Wear Test is given in Table 3. , for futher analysis MINITAB 16 software is used.

Table 2: Input parameters for Wear Test

Input Parameter	Level 1	Level 2	Level 3
% coconut fibre addition	0.5	1.0	1.5
% Glass fibre Addition	1.0	2.0	3.0
Load in Newton	10	30	50
Sliding Speed (SS) in m/s	1	2	3
Sliding Distance in m	2000		

Calculating the Signal-to-Noise ratio (□) Wear Rate has to be minimized; it is smaller- the better type of quality characteristics. Hence, S/N ratio for Mass Loss is computed from the following equation: Smaller the better $\eta = -10 \times log_{10}\left(\frac{1}{n}\Sigma y_i^2\right)$ (1)

Table 3: Results of the Wear Test

Expt.No	% 1	% 2	Load (N)	Sliding Speed (m/s)	Wear Rate (mm^3/m)
1	0.5	1	10	1	0.002

2	0.5	1	30	2	0.004
3	0.5	1	50	3	0.001
4	0.5	2	10	2	0.007
5	0.5	2	30	3	
6	0.5	2	50	1	
7	0.5	3	10	3	
8	0.5	3	30	1	0.006
9	0.5	3	50	2	0.005
10	1	1	10	2	0.002
11	1	1	30	3	0.004
12	1	1	50	1	0.001
13	1	2	10	3	0.006
14	1	2	30	1	0.008
15	1	2	50	2	0.007
16	1	3	10	1	0.004
17	1	3	30	2	0.002
18	1	3	50	3	0.001

19	1.5	1	10	3	0.004
20	1.5	1	30	1	0.002
21	1.5	1	50	2	0.001
22	1.5	2	10	1	0.004
23	1.5	2	30	2	0.002
24	1.5	2	50	3	0.001
25	1.5	3	10	2	0.003
26	1.5	3	30	3	0.004
27	1.5	3	50	1	0.007

4.7. Main Effect Plots for S/N Ratios for Wear Rate

The regression equation for the Wear Rate of Al-1-2 Metal Matrix Composites is given by:

Wear Rate = 0.64631 - (0.378127 * % 1) - (0.0451929 * %2)+ (0.00144872*Load) - (0.00146963*SS) + (0.124477 * % * %vetiver) - (0.0018006 * %2 * %2) -(1.68476e-006*Load*Load)+(0.00661419*SS*SS) +(0.0172865* %1 * %2) - (0.000837525* %1 *Load) - (0.0151358* %1 *SS) - (0.000211329* %2*Load) - (0.00854244* %2*SS) + (0.000393297*Load*SS)

S = 0.0109444 R-Sq = 99.21% R-Sq(adj) = 98.29%

The mean effect wear ration result was arrived in 98.29%. generally wear rate result above 90 % it is more sufficient. Here the result shown in table 3 and wear ratio was concluded that the result will be more sufficient for wear parameter. Hence the material used for break bad application

4.8 Fracture surface Images (SEM)

A scanning electron microscope (SEM) is a group of electron microscope that generates images of a sample by scanning it with a focused beam of electrons which are used to observe the topography and morphology of a specimen. The electrons collaborate with atoms in the sample, generating various signals that can be detected and that contain information about the sample's surface topography and composition. The analysis of fractured surfaces of impact test specimens is analyzed using SEM. The samples are taken for the SEM Analysis as they have large variations in the mechanical test results as well as variation in the composition and melting temperature. The sem image is shown in the figures.

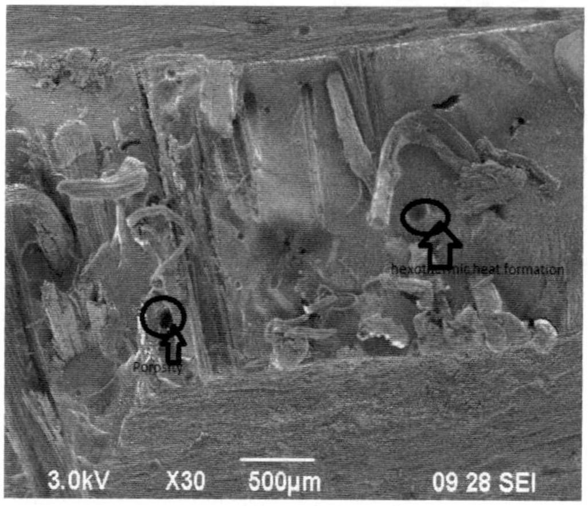

Fig 5: Tensile Fracture

Figure 10 depicts the damage that occurred due to the tensile test of the specimen. Fiber breakage occurs here due to the sheer effect which is not uniform throughout the surface because of the presence of the twisted fiber which resists each other in the opposite direction thereby attaining stability in a particular stage. The images are taken from a scanning Electron microscope. The figure 5 shown here is a ductile fracture. The porosity occurs due to exothermal heat formation. The load is given on the fiber material keep on increasing up to the elastic limit after then it will be failed.

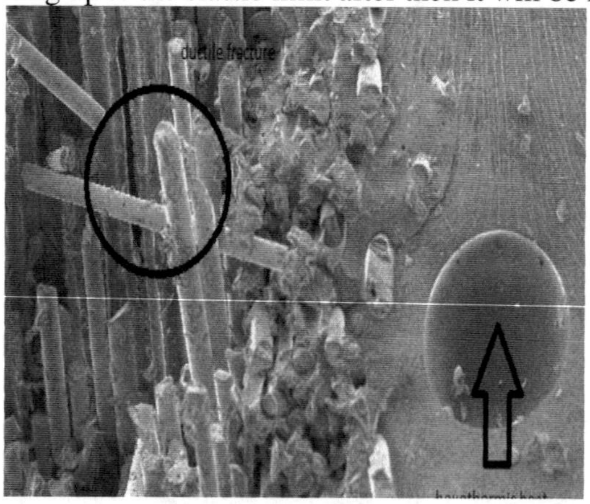

Fig 6: Double End shear Fracture

It is an internal mode failure that occurs due to the layer separation of fibers in the composite laminate. The above figure 6 shows the delaminated specimen which has its voids, blowholes. It has also been observed that during the separation of layers the adhesion of the medium has not much affected which in turn reduces the damage to the laminate. This damage has occurred due to the double shear test on the specimen. Due to the shear effect, the fiber breakage occurs which is not uniform throughout the surface because of the twisted fiber which resists each other in the opposite direction, and thereby it attains a stable state.

5. Conclusion

The tensile test results that the sample having high tensile properties rather than the other two categories in this section because of their material composition such as composites and glass fiber which is imposed on the outer layer of the laminate. The samples show a high tensile value of 4.9 Mpa with a maximum break load of 0.4 kN when compared to others. The result of the double shear test shows, the maximum value is obtained in a sample of 4.2kN. It indicates that the glass fibers have high shear properties when compared to the other two categories.The vetiver and glass fiber is made by hand layup method. The test result is shown in figures and graphs. The tensile test value results in higher UTS value in vetiver and vetiver and glass fiber. Hence that material would withstand in the above elastic region. UTS improved when the hardness value is increased hence the material withstand in a loading condition. Corresponding wear resistance is improved. The wear resistance is improved when the material withstands in corrosion wear and flexural and impact value also show the higher flexural strength and joules. The Flexural is improved when the corresponding bending strength is improved. And the impact value is improved when the ballistic load condition also improved. The fracture surface is shown in SEM images. The SEM images have shown that the Voids occur in a vetiver fiber. the vetiver and coconut fiber results in low voids and porosity. Hence It can be concluded that Vetiver and glass Fibre can be used in a Break Pad material because it has no exothermic heat, no voids, porosity, and excessive tensile strength. In the present investigation, an attempt was made in the development of the hybrid friction brake pad material to minimize the environmental and health issues and it satisfies all requirements of a brake pad material as per international- standards. From this investigation, we found an alternate material for brake bad applications. The natural fibers reinforcement material is 0% toxic. Thus the materials will not create any environmental and human health issues.

Reference

1. Pierson, Ford motor company, Emision of ammonia and amine from vechicle on the road,environment science technology 1983
2. Österle et al., Variation in the Sustained Effects of the Communities That Care Prevention System on Adolescent Smoking, Delinquency, and Violence School of Social Work, University of Washington 2001
3. Sanders The *Iras* Revised Bright Galaxy Samplethe Astronomical Journal, , 2003 October *Received 2003 February 7; accepted 2003 May 12*
4. Garg et al., Identification of brake wear particles and derivation of a quantitative tracer for brake dust at a major road Atmospheric Environment_Volume 44, Issue 2,_January 2010, Pages 141-146
5. Sanders The Human Amygdala: An Evolved System for Relevance Detection, Reviews in the neuroscience, Feb 2003
6. Garg , Identification of brake wear particles and derivation of a quantitative tracer for brake dust at a major road, Atmospheric Environment Volume 44, Issue 2, January 2010, Pages 141-146
7. **Varrica .,** Inorganic geochemistry of roadway dust from the metropolitan area of Palermo, Environment GeologyItaly **2013**
8. Sanders et al., THE *IRAS* REVISED BRIGHT GALAXY SAMPLE THE ASTRONOMICAL JOURNAL 2003
9. Hagino et al.. Revisiting the extremely fast disc wind in a gravitationally lensed quasar APM MNRAS 000, 1–12 (2017) Preprint 29 September 2018
10. Iijima Metabolite annotations based on the integration of mass spectral information us national library of medicine, year of publication 2008.
11. M. Buneaa,*, A. Cîrciumarua, etal, fiber orientation on low velocity impact behavior of the fabric reinforced hybrid compositesdoi.org/10.1016/j.compscitech.2018.11.024.

12. Brian Burks, James Middleton, Maciej Kumosa, the compare.ison of Glass fiber glass fiber reinforced composites, August 2012, Composites Science and Technology
13. Brian Burks, Maciej Kumosa, the flexural performance, dx.doi.org/10.1016/j.compscitech.2012.07.018,

14. saravana Bavan D. and Mohan Kumar G.C.A Simple hand lay-up method. Tensile and flextural test creativecommons.org/licenses/by-nc-nd/3.0/). Procedia Materials Science 5 (2014) 605 – 611

15. M. Indra Reddy a*, M. Anil Kumar a, Ch. Rama Bahadri Raju a, the tensile and flexural characterization of polymer hybrid composites, (July 29th – 30th) 2016, Materials Today: Proceedings 5 (2018) 458–462

16. Al-Hajaj, Z., Sy, B.L., Bougherara, H., Zdero, R., Impact Properties of a New Hybrid Composite Material, *Composite Structures* (2018), doi: https://doi.org/10.1016/j.compstruct.2018.10.033

17. Sivakiran.G.a, Yash Gangwal etal., Investigations on Hybrid Polymer Matrix Composite Materials, Materials Today: Proceedings 5 (2018) 7908–7914

18. Finite element analysis of fiber-reinforced polymer composites J. Naveen Mohammad Jawaid Modeling ofDamage Processes in Biocomposites,

19 Fibre-Reinforced Composites and Hybrid Composites https://doi.org/10.1016/B978-0-08-102289-4.00009-6
Azwa ZN, Yousif BF, Manalo AC, Karunasena W. A review on the degradability ofpolymeric composites based on natural fibres. Mater Des 2013;47:424e42.

21 .Mahzan S, et al. UV radiation effect towards mechanical properties of natural fibre reinforced composite material: a review. IOP Conf Ser Mater Sci Eng 2017;165:1e9.
22. Bakare FO, Ramamoorthy SK, Åkesson D, Skrifvars M. Thermo mechanical properties of bio-based composites made from a lactic acid thermoset resin and flax and flax/basalt fibre reinforcements. Composites Part A 2016.

Role of Additives for High Performance Polymer Electrolytes

Jitender Paul Sharma[1*] and Vikas Bharti[2]

[1] Department of Physics, Himachal Pradesh Technical University, Hamirpur, HP, India
[2] Department of Chemistry, Govt. Hydro Engineering College, Bilaspur, HP, India

E-mail: sharma_jp78@yahoo.co.in

Abstract. Today, the progress of polymer electrolytes with high ambient ionic conductivity, better mechanical and thermal properties, has become the main objective in the field of polymer science because of their requirements in super capacitors, rechargeable batteries, fuel cells and other electrochemical devices. Many kinds of polymer electrolytes have been developed, however main drawback is the low ionic conductivity, which is inadequate for most of the device applications. Efforts have also been made to enhance the ionic conductivity of polymer electrolytes to meet the demand for different devices. Although polymer electrolytes with high value of ionic conductivity (~10^{-3}-10^{-4} S/cm) have been developed but these also suffer from poor mechanical properties. Incorporation of plasticizers as well as nano fillers in polymer electrolytes, is of great importance for enhancing their performance in devices. In the recent years, researchers came up with an idea to make the use of ionic liquids in polymer electrolytes to address the problem of low ionic conductivity of polymer electrolytes at room temperatures as well as to keep good electrochemical and mechanical properties. The optimized ionic conductivity value of the order of ~10^{-3} S/cm has been observed by different workers at different concentrations of ionic liquids. Also, polymer gel electrolytes have been reported to possess high ionic conductivity (of the order of ~ 10^{-2} S/cm) at ambient temperature. However, despite possessing high value of ionic conductivity these gel electrolytes have poor mechanical and thermal properties. Hence, the modified polymer /gel electrolytes having optimized properties, are described in this review paper. An attempt has also been made to highlight the work of different researchers in the form of synthesis, characterization and analysis of different modified polymer electrolytes for their use in different ionic devices.

Keywords: Polymer electrolytes, Plasticizer, Filler, Ionic Conductivity, Mechanical Properties.

1. Introduction

Solid State Ionics is an interdisciplinary branch of science and technology that deals with the study of solids having high ionic conductivity value (10^{-1}-10^{-4} S/cm) which are generally known as superionic conductors or fast ion conductors. Polymer electrolytes occupy an important position in Solid State Ionics. Ion conducting polymer electrolytes are important and have received extensive attention because of their requirements in rechargeable batteries, super capacitors, fuel cells, sensors, smart windows and other electrochemical devices [1]. The research on polymer electrolytes and their use in batteries had begun in the year 1970 and continued up to now [2-6]. In past decades, researchers have focused on the implementation of polymer based high performance Li-ion batteries. High ionic conductivity, electrochemical, mechanical, thermal, interfacial and dimensional stability are the essential requirements for the materials to be used in the device applications. Since, polymer electrolytes have low ionic conductivity as well as poor mechanical properties at ambient temperature; therefore, many approaches have been adopted to enhance their conductivity to the required value applicable for devices. Common methods used to enhance the ionic conductivity of polymer electrolytes are the synthesis of plasticized as well as nanocomposite polymer /gel electrolytes. However, the enhancement in the ionic conductivity has been observed in the case of plasticized polymer electrolytes, yet deterioration of the mechanical properties of polymer electrolytes at the large extent of incorporation of plasticizer, is the existing problem. For nanocomposite polymer electrolytes, although the improvement in the ionic conductivity value is small as compared with the plasticized polymer electrolytes however, it leads to a significant modification in the mechanical properties. Recently, it has also been observed that incorporation of ionic liquids into the polymer electrolytes, has become one of the best approaches to enhance the ionic conductivity and other electrochemical properties. Optimized ionic conductivity value of the order of $\sim 10^{-3}$ S/cm has been observed in the polymer electrolytes containing different concentrations of ionic liquids by different researchers [7-11] and hence, ionic liquids have been considered as good performing dopants for polymer electrolytes. In this review article, role of additives such as plasticizer, ionic liquids and nano-filler to develop the polymer /gel electrolytes having optimum properties is described. An attempt has also been made to represent the work of various researchers towards developing the high performance polymer electrolytes.

2. Synthesis of plasticized and nanocomposite polymer/gel electrolytes

Various techniques have been implemented by the researchers to obtain free standing films of polymer electrolytes. One of the most promising techniques to obtain polymer electrolyte films, is the solution casting technique [12]. In this technique, both the salt solution and polymer solution are dissolved in a solvent and the mixture is stirred continuously with a magnetic stirrer for about couple of hours until a uniform solution is achieved. The obtained homogeneous solution is further casted in polypropylene dishes and the solvent is left to dry at room temperature for complete evaporation of solvent. The free standing films so obtained are removed from the dishes, kept in desiccator containing silica gel desiccant to dry further. Free standing and mechanically stable films are then getting ready for different experimental studies. Plasticized (polymer-salt-plasticizer or ionic liquid) and nanocomposite (polymer-salt-nano filler) polymer electrolytes are also prepared by the above method after adding stoichiometric quantities of plasticizer/ionic liquid and filler to the polymer-salt complex respectively. Polymer gel electrolytes are also prepared by the immobilizing the salt solution by adding appropriate amount of suitable polymer. Nanocomposite based polymer gel electrolytes are also developed by incorporating different concentrations of nano-filler to the obtained polymer gel electrolytes and stirring for several hours until a homogeneous and transparent solution is obtained.

3. Role of additives on the ion conducting behavior of polymer electrolytes

Complex impedance spectroscopy is often used technique to measure the electrical properties of the variety of materials. The ionic conductivity of polymer electrolytes, can be measured by sandwiching the small piece of film between two electrodes made up of stainless steel under the spring pressure. The impedance of the film, is then determined by using computer interfaced impedance analyzer in the different frequency range. Temperature dependent ionic conductivity of polymer electrolytes, can also be obtained by keeping the sample holder in temperature controlled furnace. The obtained real as well as imaginary parts of impedance (Z' and Z") are represented in Nyquist plot to measure bulk resistance (R). The ionic conductivity is then obtained from the relation [13]:

$$\sigma = \frac{G.l}{A} = \frac{l}{R.A}$$

where l is the distance between the electrodes and A is the cross-sectional area of each electrode. In this review article, the details about the electrical, structural, mechanical and thermal properties achieved by incorporating additives in the form of plasticizers or ionic liquids and fillers into the polymer electrolytes, are described as (i) plasticized polymer electrolytes and (ii) nanocomposite polymer /gel electrolytes.

3.1. Plasticized polymer electrolytes

In order to improve the ion conducting behavior of polymer electrolytes, incorporation of high dielectric constant as well as low dielectric constant plasticizers, such as propylene carbonate (PC), ethylene carbonate (EC), dimethyl sulphoxide (DMSO), dimethyl carbonate (DMC), diethyl carbonate (DEC), dimethyl formamide (DMF) and poly(ethylene glycol) (PEG) etc. into matrix polymer has been proposed [14-16]. The incorporation of high dielectric constant and high boiling temperature plasticizers dissociate the ion aggregates and un-dissociated salt into free ions [17]. The low viscosity of these plasticizers also leads to high ionic mobility that enhances the ionic conductivity [18]. The plasticizers having high dielectric constant - which leads to enhanced dissociation of salt, low viscosity - which assists the motion of charge carriers by providing low resistance conduction path in the polymer matrix, low freezing and high boiling temperature - which ensures the wide temperature range workability, have been used by number of researchers for describing their role in the enhancement of ionic conductivity of various polymer electrolytes. The addition of plasticizers to polymer electrolytes, have been reported to increase the ionic conductivity of polymer electrolytes by dissociating ion aggregates along with an improvement in amorphous phase [28-38]. The addition of plasticizer reduces glass transition temperature 'T_g', increases chain mobility and lowers the percentage crystallinity of polymers and the results obtained have been reported by using different characterization techniques such as FTIR, DSC, TGA, NMR, XRD, SEM and ECS etc. [30-31]. Finally it is concluded that the increase in ionic conductivity by introducing plasticizer, having high dielectric constant may be due to (i) increased fraction of dissociated ions, (ii) increased chain flexibility, (iii) decrease in cation-anion association (iv) cation-polymer interaction and (v) increased mobility of host polymer chains etc. Watanabe et al. [32] have studied lithium ion conducting polymer electrolytes formed by adding ethylene carbonate (EC) and lithium perchlorate ($LiClO_4$) to polymer poly(acrylonitrile) (PAN). Abraham et al. [33] also showed that the addition of PC to PAN based polymer led to a significant enhancement in the ionic conductivity such that the conductivity was higher than that of PEO based electrolytes. Park et al. [34] have observed the increase in conductivity in PAN-$LiClO_4$ based polymer electrolytes containing EC-PC as plasticizers. It has been observed that after certain concentrations of PC-EC, the conductivity begins to decrease due to the effect on the microscopic viscosity (η). Kulkarni et al. [35] synthesized PEO based plasticized polymer electrolytes by using solvent casting technique and improvement in the ionic conductivity has been observed with the incorporation of different concentrations of plasticizer DBP to PEO-$CdCl_2$ and PEO-$CdSO_4$ polymer electrolytes. Selladurai et al. [36] reported electrical, ion transport, structural and cell performance studies of polymer electrolytes containing $NaClO_3$ with the incorporation of plasticizers such as PEG, PC and DMF. Maximum ionic conductivity of the order of ~10^{-4} S/cm has been observed for PEG-PEO-$NaClO_3$-DMF system. Choudhary et al. [37] have reported enhancement in electrical conductivity of plasticized polymer electrolytes, due to increase in ions dissociation and amorphous behavior of the polymer electrolytes. They have also reported that the magnitude of conductivity depends upon the characteristics of plasticizer, plasticizer-salt concentration ratio and dielectric constant of plasticizer etc. Rajendran et al. [38] have studied PMMA, PEO and PVA based plasticized polymer electrolytes containing $LiBF_4$, $LiClO_4$ as salts and EC, DBP, DMP as plasticizers with ionic conductivity of ~10^{-4}-10^{-3} S/cm at room temperature. Subba Rao et al. [39] reported plasticized (PEO-$NaYF_4$-DMF) polymer electrolytes and it was reported that the addition of DMF to PEO-$NaYF_4$ polymer electrolytes increases the ionic conductivity, which is due to the attractive forces between the molecules of plasticizer and polymeric chain segments. This reduces the cohesive force present between polymeric chains, an increase in segmental mobility and hence, the ion conducting behavior of polymer electrolytes. Bhat et al. [40] reported the role of incorporation of plasticizer PEG on the ionic conductivity of proton conducting PEG-NH_4ClO_4 polymer electrolytes. A maximum conductivity of 5×10^{-6} S/cm has been observed for 40wt% plasticizer. The conductivity of PEO based polymer electrolytes, has been realized to increase with the addition of plasticizer, having dielectric constant greater than PEO, whereas, the conductivity has been observed to decrease by incorporating plasticizer, having dielectric constant lower than that of PEO [41]. Complexation of oxalic acid with PVdF-HFP and the effect of adding DMA to PVdF-HFP-oxalic acid polymer electrolyte has also been observed to improve the conductivity by three orders in magnitude by Chu et al. [42]. Maximum conductivity of 0.12×10^{-3} S/cm has been observed for the polymer electrolytes containing 40wt% oxalic acid with the addition of DMA. Other researchers have also made significant contribution in the development and characterization of plasticized polymer electrolytes having high ionic conductivity values along with optimized properties [43-48]. Recently, incorporation of ionic liquids as plasticizer into the polymer electrolytes has also become one of the

leading approaches to enhance the ionic conductivity and other electrochemical properties. The optimized ionic conductivity value of the order of ~10^{-3} S/cm has been observed at different ionic liquid concentrations by different researchers [7-11] due to which, these ionic liquids have been considered as good performing dopants for polymer electrolytes.
Since, plasticized polymer electrolytes exhibit relatively higher ionic conductivity, but a poor mechanical property is still the existing problem [28]. Therefore, an efficient and effective method is required for the improvement of mechanical properties of plasticized polymer electrolytes. On the basis of above, the authors have decided to synthesize and characterize high performance polymer electrolytes with reduced crystalline properties and enhanced amorphous properties for future perspectives.

3.2. Nanocomposite polymer /gel electrolytes

(i) Nanocomposite Polymer Electrolytes (in film form)

One of the steps to improve room temperature ionic conductivity as well as mechanical properties of polymer electrolytes is the addition of fillers. Use of fillers such as silica, zeolites, oxides as additives in polymer electrolytes have shown an enhancement in ionic conductivity along with better mechanical stability [27, 49-61]. These nanocomposite polymer electrolytes have certain advantages over organic solvent based polymer electrolytes in the form of mechanical properties, ionic conductivity and electrode-electrolyte interfacial stability [62-63]. The enhancement in Li^+ ion conductivity by three orders of magnitude was because of the dispersion of filler Al_2O_3 in Lithium Iodide as firstly noted by Liang [64] in 1973. In 1982, Weston and Steele [65] reported the enhancement in the mechanical stability of PEO-$LiClO_4$ electrolytes using α-Al_2O_3 particles with grain size 40 μm. In earlier studies on mixed-phase polymer electrolytes consisting of NASICON [66], β-alumina [67], it has been shown that fillers do not show any enhancing effect to the ion conducting behavior of mixed phase systems. However, Skaarup et al. [68] reported the decrease in conductivity for composite polymer electrolytes containing high amounts of conducting fillers. Similar results have also been investigated by Stevens and Mellander [69] on the conductivity enhancement of mixed-phase electrolytes by introducing conducting fillers. Scrosati et al. [70] have reported the effect of dispersion of γ-$LiAlO_2$ to PEO based polymer electrolytes. No significant improvement in the ionic conductivity as well as electrochemical properties has been observed. Morales et al. [71] have also recorded that the ionic conductivity of the system decreases by dispersing γ-$LiAlO_2$ at low concentrations. An improvement in the ionic conductivity was recorded at higher concentrations of γ-$LiAlO_2$. In composite polymer electrolytes, surface interactions have been assumed to be answerable for enhancement in ionic conductivity as reported by Maier et al. [72]. Researchers have proposed different models to depict the variations in the electrical and mechanical properties of composite polymer electrolytes with polymer, salt and filler [73]. Croce et al. [74] have proposed a model for the effect of filler in polymer matrix to reduce the crystallization as well as to promote the specific interaction between surface groups and PEO segments. Greenbaum et al. [75] have studied PEO based composite polymer electrolytes in terms of electrochemical and solid state NMR characterizations. It has also been recorded that the Li^+ ion diffusivity can be enhanced without any modification in segmental mobility of polymeric chains. Walls et al. [62] have reported high conductivity at room temperature by dispersing fumed silica to poly(ethylene glycol) dimethyl ether (PEGDME) containing lithium salt. However, the conductivity showed a tendency to decrease above 20wt% fumed silica. Dissanayake et al. [76] have reported that the presence of nanoporous Al_2O_3 particles having four different surface groups in PEO-LiTFSI polymer electrolytes enhances the ionic conductivity depending upon the nature of the filler surface group. Forsyth et al. [77] have studied the phase change in PEO based polymer electrolytes with the addition of TiO_2. They found that the polymer segmental motion was enhanced at low filler concentration but the enhancement diminishes at high filler concentration. Capuano et al. [70] have reported an enhancement in the ionic conductivity with the dispersion of 10wt% Al_2O_3 to PEO-$LiClO_4$. A significant enhancement in the interfacial stability has also been recorded due to the reduced crystallinity of polymer electrolytes. Kumar et al. [78] reported that nano fillers are more compatible to reduce the interfacial resistance than fillers with micron size. Generally smaller the particle size, larger the conductivity enhancement [84]. Bhat et al. [79] have noted the significant improvement in Li^+ ion conductivity in PEO based polymer electrolytes containing $LiClO_4$ after dispersing different concentrations of δ-Al_2O_3 nanoparticles. It was concluded that the decrease in percentage crystallinity and activation energy may be due to increase in mobility in lithium ions. The maximum conductivity of 4.5×10^{-6} S/cm has been observed with the addition of filler. Morita et al. [80] have investigated the role of addition of ceramic fillers on the ion transport properties of poly(ethylene oxide)-poly(methacrylate) (PEO-PMA) based polymer electrolytes, containing magnesium salt $Mg[(CF_3SO_2)_2N]_2$. The addition of ceramic filler has also been observed to improve the ionic conductivity due to the enhancement in ionic mobility. Maximum conductivity of $\sim 10^{-5}$ S/cm at 60^oC has been observed for composite polymer electrolytes. The dispersion of ceramic filler has also been observed to influence the thermal properties of polymer electrolytes. Khan et al. [81] have reported PEG based composite polymer electrolytes developed by dispersing hydrophobic fumed silica in PEG-Li salt solution. Along with an improvement in mechanical stability of polymer electrolytes, the ionic conductivity of the order of 10^{-3} S/cm has also been recorded. From the literature, two conductivity maxima at two different filler concentrations have been observed in case of composite polymer electrolytes due to the mobility of both the mobile ions [82-83]. Fan et al. [84] have also reported two conductivity maxima for composite PEO based polymer electrolytes containing $LiClO_4$ and silane modified SiO_2. The enhancement in the ionic conductivity observed due to the dispersion of ceramic filler, has been described by an effective medium theory (EMT) and existence of high conducting layer at electrolyte/filler interface. Composite solid polymer electrolytes with improved mechanical properties and electrochemical stability have been developed by incorporating MnO_2 nano-sheets for lihium metal batteries by Shujiang et al. [3]. PEO/MnO_2 composite solid polymer electrolyte exhibited higher ionic conductivity along with better tensile strength than PEO solid electrolytes. Ion transport properties of novel blended PEO-PVP based polymer electrolytes for the sodium ion rechargeable batteries have been prepared via standard solution cast technique by Sharma et al. [4]. To obtain more information about the morphology and structural changes in films, these polymer electrolytes have also been studied by using different characterization techniques. Still the main challenges in the commercialization of Li-ion battery are the low ionic conductivity and poor mechanical properties. In order to improve the overall performance of polymer-based composite electrolytes, Liu et al. [5] have obtained the ionic conductivity of the order of 10^{-4} S/cm along with electrochemical window of 0–5.0 V. Saito et al. [6] have fabricated cross-linked polymer electrolytic membranes containing branched poly (ethyleneimine), (PEI) and poly(ethylene oxide), (PEO). The optimized polymeric membranes have shown excellent ionic conductivity value of the order of 10^{-3} S/cm along with many potential applications in electrochemical storage devices.

From the above overview of literature, it has been observed that no doubt, many studies have been recorded in case of nanocomposite polymer electrolytes, yet several issues are still unresolved. The purpose of this review is to provide an idea which will help to access the influence of dispersion of fillers as well as to guide the readers towards their best efforts in future research work. Therefore, authors have made attention to obtain polymer electrolytes having optimized ionic conductivity as well as improved mechanical and thermal properties over a wide temperature range.

(ii) Nano-dispersed polymer gel electrolytes

Polymer gel electrolytes are generally obtained by immobilizing ionic salt solution with suitable polymer matrix to get a hybrid mixture [85]. In polymer gel electrolytes, free ions are mainly provided by doping ionic salt. The solvent used acts as a conducting medium and helps to solvate the ionic salt whereas; polymer provides the mechanical stability by modifying the viscosity of polymer electrolytes [86]. Different polymers used in the preparation of polymer gel electrolytes are poly(vinylidene fluoride) PVdF [87-88], poly(methylmethacrylate) (PMMA) [89-90], poly(acrylonitrile) (PAN) [91], poly(vinyledenefluoride-hexafluoropropylene) (PVdF-HFP) [92] etc. Singh et al. [93] have reported transport properties of different PMMA based polymer gel electrolytes containing $LiCF_3SO_3$ salt having ionic conductivity of $\sim 10^{-2}$ S/cm at room temperature. Scrosati et al. [94] have studied lithium ion conducting and proton conducting polymer gel electrolytes and found their suitability in Li^+ ion batteries and fuel cells respectively. No doubt, initially the work was concentrated on gel electrolytes containing various lithium salts mainly, however, proton conducting polymer gel electrolytes have also been described due to their suitability in many devices [95]. Protons conducting polymer gel electrolytes containing H_3PO_4, H_2SO_4, HCl [96], hydroxy benzoic and aliphatic dicarboxylic acids [97] have been recorded with conductivity higher than the corresponding liquid electrolytes. Arora et al. [98] have described PMMA based polymer gel electrolytes containing different acids with PC, EC and DMF as solvents. The modification in ionic conductivity with the addition of PMMA has been described on the basis of "Breathing polymer chain model". In accordance with this model, folding and unfolding of the polymer chain occurs due to the breathing which implies the pressure fluctuations at microscopic level [98]. This also helps to dissociate ion aggregates as well as un-dissociated salt which results in an enhancement in ionic conductivity value. Some of the recent publications by different workers on the polymer gel electrolytes have also been identified [98-101]. From the above discussion, it has been concluded that polymer gel electrolytes have been observed to possess high ionic conductivity, but their mechanical and thermal stability still need further improvement. Data of some available patents on different polymer electrolytes has also been shown in Table 1.

Table 1. Data of some available patents on polymer electrolytes

Sr. No.	Inventor	Polymer Electrolyte (Title)	Conductivity order (S/cm)	Patent Number (Year)
1	K. M. Abraham M. Alamgir	Mixed polymer electrolyte and mixed polymer electrolyte battery	10^{-5}	US5166009A (1992)
2	B. Kumar L. G. Scanlon Jr. J.D. Schaffer	Polymer-ceramic composite electrolytes	10^{-5} -10^{-3}	US5695873A (1997)
3	M. Zafar A. Munshi	Solid polymer electrolytes	10^{-3} - 10^{-4}	CA2382118A1 (2001)
4	W. Smith J. McCloskey	Polycarbonate oligomers and polymers for use in electrolytes	10^{-3}	US20020168575A1 (2002)
5	Y. Kang C. Lee S. K. Kim	Polyalkylene oxide polymer composition for solid polymer electrolytes	10^{-4}	US20030044688A1 (2003)
6	M. Singh I. Gur H. B. Eitouni N. P. Balsara	Multiple electrolyte electrochemical cells Images	10^{-5}	US20110281173A1 (2011)
7	M. Singh I. Gur H. B. Eitouni N. P. Balsara	Solid electrolyte material manufacturable by polymer processing methods	10^{-5}	8268197 (2012)
8	Y. S. Kim D. S. Kim K. S. Lee	Anion exchange polymer electrolytes	10^{-3}	US8492049B2 (2013)
9	B. Hsieh H. B. Eitouni M. Singh	Polymer electrolyte materials based on block copolymers	10^{-5}	US20120029099A1 (2014)
10	M.A. Zimmerman	Solid ionically conducting polymer material	10^{-4}	WO2016182884A1 (2016)
11	P. Kofinas A. Ghosh	Polymer Solid Electrolyte for Flexible Batteries	10^{-4}	US20100255383A1 (2016)

12	M.A. Zimmerman	Solid electrolyte high energy battery	10^{-4}	9819053 (2017)
13	S. L. Wunder P. R. Chinnam	Solid polymer electrolyte composition for lithium ion battery	10^{-4}-10^{-3}	US9680182B2 (2017)
14	D. W Ko D. K. Yang E. K. Park J. H. Chae	Polymer Electrolyte Having Multi-Layer Structure, And All-Solid Battery Comprising Same	10^{-4}	EP3285324A1 (2018)
15	J. Nanda R. E. Ruther F. M. Delnick C. N. Sun	Electrolytes for supercapacitors	10^{-3}	10102982 (2018)
16	R. C. Pratt X. L. Wang S. Lam H. B. Eitouni	Fluorinated alkali ion electrolytes with urethane groups	10^{-5}	9923245 (2018)
17	Y. Shuting G. Zhiwen W. Qiuxian Y. Hongyun W. Zhaoxiang C. Liquan	A kind of composition polymer solid electrolyte membrane and preparation method thereof	10^{-3}	CN109546211A (2019)
18	L. Ronghua W. Jinxiang S. Jun	Polymer dielectric film and preparation method thereof and lithium ion battery	10^{-6} - 10^{-4}	CN110117417A (2019)
19	C. Qing Z. Juntang	A kind of dynamic lithium battery microcapsules thin-film ceramics solid electrolyte and preparation method	10^{-4} - 10^{-3}	CN108134132B (2019)
20	J. Y. Sanchez C. Iojoiu Y. Molmeret C. Antonelli	Ionically Conductive Material For Electrochemical Generator And Production Methods	10^{-5} - 10^{-3}	20190229315 (2019)

In order to overcome limitations attained by polymer gel electrolytes, the effect of dispersion of filler to polymer gel electrolytes will also be one promising line of investigation to develop nano-dispersed polymer gel electrolytes, having high ionic conductivity values, good mechanical as well as thermal properties. Still little work has been done on the addition of nanoparticles to the gel electrolytes. Therefore, authors would also like to develop and study nano dispersed polymer gel electrolytes, especially by making the use of biodegradable polymer, in order to replace traditional polymer electrolytes to optimize their properties as earlier as possible.

4. Summary

Generally, polymer electrolytes have been observed to possess low ionic conductivity, poor mechanical as well as thermal property due to which there is a need to synthesize high performing polymer electrolytes. Although, incorporation of plasticizer has increased the ionic conductivity value as discussed yet improper concentration may deteriorate the mechanical properties of polymer electrolytes. Nanocomposite polymer electrolytes have shown the significant improvement in mechanical as well as thermal properties of polymer electrolytes; however ion transport properties still need further improvement. To address the problem of low ionic conductivity of polymer electrolytes, the addition of ionic liquids as plasticizer in polymer electrolytes has also been considered as one of the promising approaches. Therefore, in order to overcome the limitations of polymer electrolytes having low ionic conductivity values, the addition of additives in the combined form such as plasticizer/ionic liquid and nano filler will be one of the leading approaches. Since, polymer gel electrolytes have been reported to possess high ionic conductivity at room temperature. However, despite possessing high ionic conductivity values of the order of ~ 10^{-2} S/cm, these gel electrolytes have poor mechanical and thermal properties. The dispersion of nano filler to polymer gel electrolytes to form nano-dispersed gels/gel electrolytes with improved ion transport properties, will also be an interesting study. Hence, the research and development efforts of the authors to study the combined effect of additives on the ion conducting behavior of environmentally friendly, non-toxic and low cost polymer electrolytes, will be the next step.

5. References

[1] McCallum J.R., Vincent C.A. 1991 *Polymer Electrolyte Reviews-2*, Elsevier, Amsterdam 1989, p 338.
[2] Armand, M.B., Chabagno, J.M. and Duclot, M.J. 1979. Fast ion transport in solids. *Electrodes and Electrolytes*, p 131.
[3] Yuhan Li, Zongjie S., Dongyu Li., Yiyang G., Yuankun W., Huaitian Bu, Mingtao Li, Yanfeng Z., Guoxin G., Shujiang D. 2020 *J . Mater. Chem . A* **8**, 2021-2032.
[4] Pritam, Arya A., Sharma A.L. 2019 *J. Mater. Sci.* **54**, 7131-7155.
[5] Yao P., Yu H., Ding Z., Liu Y., Lu J., Lavorgna M., Wu J., Liu X. 2019 *Front. Chem.* **8(7)**, 522.
[6] Lehmann M.L., Yang G., Nanda J., Saito T. 2020 *J. Electrochem. Soc.* **167**, 070539.
[7] Sekhon S. S., Lalia B. S., Park J.S., Kim C.S., Yamada K. 2006 *J Mater Chem* **16**, 2256-2265.
[8] Kar, M., Tutusaus, O., MacFarlane, D.R., Mohtadi, R. 2019 *Energy Environ Sci* **12**, 566-571.
[9] Yang G, Song Y, Wang Q, Zhang L, Deng L. 2020 *Materials & Design* **190**, 108563.
[10] González F, Garcia-Calvo O, Tiemblo P, García N, Fedeli E, Thieu T, Urdampilleta I, Kvasha A. 2020 *J Electrochem. Soc* **167(7)**, 070519.
[11] Singh P. K., Jadhav N, Mishra S. K., Singh U. P., Bhattacharya B. 2010 *Ionics* **16(7)**, 645-648.
[12] Kumar M., Sekhon S.S. 2002 *Euro. Polym. J.* **38(7)**, 1297-1304.
[13] Atkins P. W. 1998 *Introduction to Physical Chemistry*. Freeman, New York, p 999.
[14] Appetecchi G. B., Croce F., Scrosati B. 1995 *Electrochim Acta* **40(8)**, 991-997.
[15] Alangmir M., Abraham K.M. 1994 *Lithium batteries-new materials, developments and perspectives.* ed G. Pistoia, Elesevier, Amsterdam.
[16] Forsyth M., Meakin P., MacFarlane D. R., Bulmer G., Ried V. 1994 *J. Mater. Chem.* **4**, 1149-1151.
[17] Bishop A.G., MacFarlane D.R., McNaughton D., Forsyth M. 1996 *Solid State Ionics* **85(1-4)**, 129-135.
[18] Southall J.P., Hubbard H.V.St.A., Johnson S.F., Rogers V., Davies G.R., McIntyre J.E., Ward I.M. 1996 *Solid State Ionics* **85**, 51-60.
[19] Kumar M., Sekhon S.S. 2002 *Euro Polym J* **38**, 1297-1304.
[20] Tsutsumi H., Matsuo A., Onimura K., Oishi T. 1998 *Electrochem Solid State Lett* **1(6)**, 244-245.
[21] Hamsan M. H., Aziz S.B., Shukur M.F.A., Kadir M.F.Z. 2019 *Ionics* **25 (14-16)**, 1-14.
[22] Wieczorek W., Raducha D., Zalewska A., Stevens J. R. 1998 *J Phys Chem B* **102(44)**, 8725-8731.
[23] Xue Z., He D., Xie X. 2015 *J Mater Chem A* **3(38)**, 19218-19253.
[24] Das S., Ghosh A. 2017 *Phys Chem B* **121(21)**, 5422-5432.
[25] Kuo C. W., Li W. B., Chen P.R., Liao J.W., Tseng C.G., Wu T.Y. 2013 *Int J Electrochem Sci* **8(4)**, 5007-5021.
[26] Rajendran S., Sivakumar M., Subadevi R. 2004 *Mater Lett* **58**, 641-649.
[27] Gray F.M. 1991*Solid Polymer Electrolytes: Fundamental and Technological Applications.* VCH, New York.
[28] MacFarlane D.R., Sun J., Meakin P., Fasoulopoulos P., Hey J., Forsyth M. 1995 *Electrochim. Acta* **40 (13-14)**, 2131-2136.
[29] Berthier C., Gorecki W., Minier M., Armand M.B., Chabagno J.M., Rigaud P. 1983 *Solid State Ionics* **11(1)**, 91-95.
[30] Cheradame H., Le Nest J.F. 1987 *Polymer Electrolyte Reviews-I*, Elsevier Applied Science, London.
[31] Bishop A.G., MacFarlane D.R., McNaughton D., Forsyth M. 1996 *Solid State Ionics* **85**, 129-135.
[32] Watanabe M., Kanba M., Nagaoka K., Shinohara I. 1983 *J. Polym. Sci.: Polym. Phys.* **21**, 939-948.
[33] Abraham K.M., Alamgir M. 1990 *J. Electrochem. Soc.* **136**, 1657-1658.
[34] Park U.S., Hong Y.J., Oh S.M. 1996 *Electrochim. Acta* **41**, 849-855.

[35] Mitra S., Kulkarni A.R. 2002 *Solid State Ionics* **154-155**, 37.
[36] Chandrasekaran R., Selladurai S. J. 2001 *Solid State Electrochem.* **5**, 355-361.
[37] Pradhan D.K., Samantaray B.K., Choudhary R.N.P., Thakur A.K.. 2005 *J. Power Sources* 139, 384-393.
[38] Rajendran S., Siva Kumar M., Subadevi R. S. 2003 *J. Power Sources* **124**, 225-230.
[39] Reddy M.J., Sreekanth T., Subba Rao U. V. 1999 *Solid State Ionics* **126** (1-2), 55-63.
[40] Binesh N., Bhat S. V. 1999 *Solid State Ionics* **122**, 291-299.
[41] Sekhon S.S., Kumar M. 2002 *Solid State Ionics: Trends in the New Millennium.* eds. Chowdari B. V. R., Prabaharan S.R.S., Yahaya M., Talib I.A. World Scientific, Singapore p 377.
[42] Missan H. P. S., Chu P. P., Sekhon S. S. 2006 *J. Power Sources* **158**, 1472-1479.
[43] Sharma J. P., Singh V. 2020 *High Performance Polymers* **32(2)**, 142-150.
[44] Kuo C., Li W., Chen P., Liao J., Tseng C., Wu T. 2013 *Inter. J. of Electrochem. Sci.* **8**, 5007-5021.
[45] Noor M. M., Careem M. A., Majid S. R., Arof A. K.. 2011 *Mater. Res. Innov.*, **15(2)**, 157-160.
[46] Rahman M. Y. A., Ahmad A., Lee T. K., Farina Y., Dahlan H. M.2011 *Mater. Sci. and Appl.*, **2**, 818-826.
[47] Jaafar N.K., Lepit A, Aini N.A, Ali, A.M.M, Saat A, Yahya M.Z.A. 2014 *Int. J. Electrochem. Sci.*, **9**, 821-829.
[48] Ahmed H.T., Jalal V.J., Tahir D.A., Mohamad A.H., Abdullah O.G.2019 *Results in Phys*;**15**, 102735.
[49] Scrosati B. ed. 1993 *Application of Electroactive Polymers*. Chapman and Hall, London.
[50] Abraham K. M., Koch V. R., Blakley T. J. 2000 *J. Electrochem. Soc.* **147(4)**, 1251-1256.
[51] Rosli, N.H.A., Chan C.H, Subban R.H.Y, Winie T. Phys. 2012 *Procedia* **25**, 215–220.
[52] Jayakumar R, Prabaharan M, Nair S.V, Tamura H. 2010 *Biotechnol. Adv.***28**, 142–150.
[53] Day A., Karan S., De S.K. 2009 *Solid State Chomm.* **149**, 1282-1287.
[54] Radha K.P. Der. 2017 *Pharma Chemica* **9(4)**, 43-47.
[55] Li, J., Miao, D., Yang, R., Qu, L. and Harrington, P. B. 2014 *Electrochimica Acta*, **125**, 1-8.
[56] Ramesh S., Wen L.C. 2010 *Ionics* **16**, 255-262.
[57] Senthil R.A. 2015 *Mater Sci Forum,* **832**, 72-83.
[58] Mohamad A.H., Saeed S.R., Abdullah O.G. 2019 *Mater Res Express* **6**, 115332.
[59] Rani, M.S.A., Ahmad, A., Mohamed, N.S. 2017 *Ionics* **24**, 807-814.
[60] Ponnamma D., Erturk A., Parangusan H., Deshmukh K., Ahamed M.B., Al-Maadeed M.A.A. 2018 *Emergent Mater* **1**, 55-65.
[61] Solarajan A.K., Murugadoss V., Angaiah S. 2017 *J. Appl. Polym. Sci.* **134**.
[62] Walls H. J., Zhou J., Yerian J. A., Fedkiw P. S., Khan S. S., Stowe M. K., Baker G. L. 2000 *J. Power Sources* **89** **(2)**, 156-162.
[63] Florjanczyk Z., Marcinek M., Wieczorek W., Langwald N. 2004 *Polish J. Chem.* **78**, 1279-1304.
[64] Liang C.C. 1973 *J. Electrochem. Soc.* **120**, 1289-1292.
[65] Weston J. E., Steele B. C. 1982 *Solid State Ionics* **7**, 75-79.
[66] Plocharski J., Wieczorek W. 1988 *Solid State Ionics* **28-30**, 979.
[67] Wieczorek W., Such K., Przyluski J., Florjanczyk Z. 1991 *Synth. Met.* **45**, 373-378.
[68] Skaarup S., West K., Julian P.M., Thomas D. M. 1990 *Solid State Ionics* **40-41**, 1021-1024.
[69] Stevens J.R., Mellander B. E. 1986 *Solid State Ionics* **21**, 203-206.
[70] Capuano F., Croce F., Scrosati B. 1991 *J. Electrochem. Soc.* **20**, 1918-1922.
[71] Morales E., Acosta J. L. 1998 *Solid State Ionics* **111**, 109-115.
[72] Maier J. 1989 *Superionic Solids and Solid Electrolytes-Recent Trends*, eds A. L. Laskar, S. Chandra, Academic Press, New York, p 137.
[73] Hashmi S.A., Thakur A. K., Upadhyaya H. M. 1998 *Euro. Polym. J.* **34**, 1277-1282.
[74] Croce F., Persi L., Scrosati B., Fiory F. S., Plichta E., Hendrikson M. A. 2001 *Electrochim. Acta* **46(16)**, 2457-2461.
[75] Cheung I.W, Chin K.B., Greene E.R., Smart M.C., Abbrent S., Greenbaum , Prakash G.K.S., Surampudi S. 2003 *Electrochim. Acta* **48**, 2149-2156.
[76] Jayathilaka P.A.R.D., Dissanayake M.A.K.L., Abinsson I., Mellander B. E. 2002 *Electrochim. Acta* **47**, 3257-3268.
[77] Forsyth M., MacFarlane D.R., Best A., Adebahr J., Jacobbson P., Hill A.J. 2002 *Solid State Ionics* **147**, 203-211.
[78] Kumar B., Scanlon L.G., Spry R.J. 2001 *J. Power Sources* **96**, 337-342.
[79] Joykumar T., Bhat S.V. 2004 *J. Power Sources* **129**, 280-287.
[80] Morita M., Noborio H., Yoshimoto N., Ishikawa M. 2006 *Solid State Ionics* **177**, 715.
[81] Raghavan S.R., Riley M.W., Fedkiw P.S., Khan S.A. 1998 *Chem. Mater.* **10**, 244-251.

[82] Thakur A.K., Upadyyaya H.M., Hashmi S.A., Verma A.L. 1998 *Solid State Ionics: Science and Technology*. eds B.V.R. Chowdari, K. Lal, S.A. Agnihotry, N. Khare, S.S. Sekhon, P.C. Srivastava, S. Chandra, World Scientific, Singapore, p 229.
[83] Lakshmi N., Chandra S. 2001 *Phys. Stat. Sol. (a)* **186**, 383-399.
[84] Fan L., Nan C. W., Zhao S. 2003 *Solid State Ionics* **164(1-2)**, 81-86.
[85] Feuillade G., Perche Ph. 1975 *J. Appl. Electrochem.* **5**, 63-69.
[86] Sekhon S. S., Pradeep, Agnihotry S. A. 1998 *Solid State Ionics: Science and Technology*. eds. B.V.R. Chowdari, K. Lal, S.A. Agnihotry, N. Khare, S.S. Sekhon, P.C. Srivastava, S. Chandra, Singapore, World Scientific, p 217.
[87] Sakia D., Kumar A. 2004 *Electrochim. Acta* **49**, 2581-2589.
[88] Saito Y., Capigila C., Yamamoto H., Mustarelli P. 2000 *J. Elecrochem. Soc.* **147(5)**, 1645.
[89] Vondrak J., Reiter J., Velicka J., Sedlarikova M. 2004 *Solid State Ionics* **170(1-2)**, 79-82.
[90] Appetecchi G.B., Romagnoli P., Scrosati B. 2001 *Electrochem. Commu.* **3**, 281-284.
[91] Croce F., Gerace F., Dautzenberg G., Passerini S., Appetecchi G.B., Scrosati B. 1994 *Electrochim. Acta* **39(14)**, 2187-2194.
[92] Kim S.H., Choi J.K., Bae Y.C. 2001 J. *Appl. Polym. Sci.* **81**, 948-956.
[93] Singh B., Kumar R., Sekhon S.S. 2005 *Solid State Ionics* **176**, 1577-1583.
[94] Ciuffa F., Croce F., Epifanio A. D., Panero S., Scrosati B. 2004 *J. Power Sources* **127**, 53-57.
[95] Colomban Ph. ed 1992 *Proton Conductors: Solid membranes and gel-materials and devices.* Cambridge: University Press.
[96] Wieczorek W., Zukowska G., Borkowska R., Chung S.H., Greenbaum S. 2001 *Electrochim. Acta* **46**, 1427-1438.
[97] Chandra S., Sekhon S.S., Arora N. 2000 *Ionics* **6**, 112-118.
[98] Deraman S.K., Mohamed N.S., Subban R.H.Y. 2013 *Sains Malaysiana* **42(4)**, 475-479.
[99] Pandey, K., Dwivedi, M.M., Asthana, N., Singh, M. and Agrawal, S.L. 2011 *Mat. Sci. and Appl.,* **2**, 721-728.
[100]Cavus S., Durgun E. 2016 *Acta Physica Polonica A*, **129(4)**, 621-624.
[101]Ravi M., Song S., Wang J. 2016 *Inter. J. Ionics* **22(5)**, 661-670.

Acknowledgement
Authors are highly thankful to the Himachal Pradesh Technical University, Hamirpur (H.P.) for providing financial assistance under the sanctioned Competitive Research Fund of Collaborative Research and Innovation Programme TEQIP-III vide office letter no. HIMTU-3 (Acad) M-CR/2019/718-722, 2019 with topic entitled "Development of highly conductive Polymer electrolytic membranes for potential applications".

IOP Conf. Series: Materials Science and Engineering **961** (2020) 012006 doi:10.1088/1757-899X/961/1/012006

Studies on application of image processing in various fields: An overview

T Prabaharan[1], P Periasamy[2],V Mugendiran[3],Ramanan[4]

[1]Research Scholar, St. Peter's Institute of Higher Education and Research, Chennai, India
[2]Professor, Department of Mechanical Engineering, St. Peter's Institute of Higher Education and Research, Chennai, India
[3]Professor, Department of Production Technology, Madras Institute of Technology, Anna University, Chennai, India
[4] Sr Engineer Synce Engineering Service, Chennai

Email:prabaezhile1977@gmail.com

AbstractQuality inspection and evaluation is a vital role in producing quality products in a shorter span. Computer-aided estimation of quality products in different fields of engineering is a constructive advancement happening. Image processing is one of the most promising areas, which is applied for quality inspection of products where the challenging task lies in recognition of the object and feature extraction. This paper made attempt to provide an overview of the application of image processing and their methodology with few algorithms that have been used in different fields of engineering which falls under three important phases: acquisition of images, the region of interest and identification of defects. This paper concentrates on applications like construction, fluid flow, thermal imaging, medical industries, fruit and vegetable industry, rock carvings, and other applications. Applying image processing different fields leads to quality products by a qualitative process which leads to a reduction in inspection time and cost involved.

Keywords: Image processing; Machine vision; Data acquisition; Strain measurement; Thermal; Fluid flow;

1. Introduction

Image processing technique can be used for processing images, 3d models, printouts and to obtain the required data from the images. Researchers use a broad range of basic procedure of image interpretation while adopting analog visual techniques. This type of image processing is just restricted within the area of knowledge of the analyst. So analysts may apply a blend of personal knowledge and data in image processing. In digital image processing, computer based algorithms are developed to perform image processing technique. Considering the advantages of digital image processing against analog image processing and due to huge number of algorithms available that can be used with the input data. In digital image processing, few problems during processing such as noise creation, signal distortion etc., can be minimized and removed during preprocessing technique called signal processing. In late 2000, due to the advancement that happened in digital image processing with aid of computers has become the emerging form of image processing which is more versatile, and also the cheapest one. Image processing has strong relation with computer vision and computer graphics. The

IOP Conf. Series: Materials Science and Engineering **961** (2020) 012006 doi:10.1088/1757-899X/961/1/012006

following steps describe the procedure for image processing: Hallucination (identifying the hidden objects), Image restoration and sharpening (for creating sharpened image), Image repossession (search for the area of interest), Measurement of pattern (calculating the color range of objects) and Image acknowledgment (differentiating the region of interest). In this study, a review on digital image processing, applied in various field has been given with suitable algorithms.

2. Image processing in different fields

It can be noticed that substantial amount of investments has been going on civil infrastructures since past few decades. To assure the safety of the civilians, the priority has to be given to maintenance and the interventions should be defined to reduce both environmental impacts and costs. Due to climatic changes, the former leads to different new maintenance strategies. To achieve quick and reliable diagnostics, focused solution has to be developed to maintain the structures' robust. Thus the developed solution ought to ensure be effective, reliable and economical forever. The availability of the digital and optical equipment's gained importance in structural assessment. Currently, in construction sector to perform land survey, structural damage monitoring, structural health assessment, deformation and damage study [2]. Terrestrial Laser Scanning (TLS) method has been widely adopted. By means of 3D images of structures, this accurately collects both qualitative and quantitative information. For characterization and monitoring, Terrestrial Photogrammetry (TP) has been widely adopted [3]. Photogrammetry allows cost effective high-resolution 3D structural imaging systems. [4] It is not possible to characterize fluid flow pattern with high velocity. Image processing technique is most widely adopted to visualize and characterize complicated 3D fluid flow to acquire clear image of the physical phenomena for further processing. The foremost interests for researchers in fluid flow are pattern formation and flow structures, by analysis the phenomena through acquired images. Much more study has been carried out on jet origination, propagation, hypersonic jet flows, flow structures and patterns and also morphological flow properties, and their emission, by means of image processing [10]. Severe heat generation occurs in case of friction during machining, generation of electricity and flow of electricity. Even in building construction too, heat loss occurs due to poor wall finishing, roof performance and poor insulation. If the insulation provided is too weak then it loses its effectiveness, and can lead to even worst damage to the structure and also to the interior of the building. For example, motors running at a same load conditions continuously, should be inspected for change in thermal properties, so that early detection may avoid catastrophic failure[24]. The diagnosis of various human diseases can be completed through common available medical tests. Biomedical images are acquired from living beings and they are used for clinical diagnostics, disease treatment and continuous monitoring. Medical imaging can be applied to observe and study thefunction and behavior of internal organs without the need of surgery. Computer Tomography (CT), Magnetic Resonance Imaging (MRI) and X-ray inspection are the few commonly used imaging techniques in contemporary medical field. Virtual reality and augmented reality are the innovation that has been applied to save and improve the quality of life. Several sensors can be also being employed to study and monitor the health condition of the human like blood pressure, body temperature, air respiratory, glucose level, skin perspiration etc., however molecular analysis study using microscopic images can be used to recognize the symptom of the diseases. Study of strain developed on a specimen during different types of loading, say tensile, compression, shear, bending, twisting, etc., and plays an important role in prediction of life of the specimen. Even fluctuating stresses causes a major problem on components subjected to dynamic loads. The prediction of strain distribution and evaluation of strain localization behavior before the failure of the components is the major area that has to be concentrated. In this method, a random pattern is etched on the top layer of the sample, in order to determine the spatial displacement of the pattern on the sample due to different loading conditions using digital correlation calculation. The deviation of the electrochemically or laser etched grid patterns on the sample has been analyzed by developing computer based algorithms [11].

IOP Conf. Series: Materials Science and Engineering **961** (2020) 012006 doi:10.1088/1757-899X/961/1/012006

3. Algorithms used in Image Processing

3.1Image processing in construction industry

Image processing can be seen as automated system to perform the health monitoring and evaluation system to assess the damage occurring in concretes and structures due to natural calamities [5]. This deals with the characterization of crack patterns, measurement of strain fields subjected to different loads. The automation in characterization leads to measuring and comparing the entire crack length over a period of time, eliminating the human error with high accuracy [6]. The main drawback is the resolution quality of the image obtained which has a vital role in image acquisition [7]. Algorithms has been developed to detect a crack developed on the surface of the concrete structure based on image acquired using an automated robot [1,8,9]. The processes involved in image acquisition and image processing on concrete structures are shown as a flow chart in figure 3.

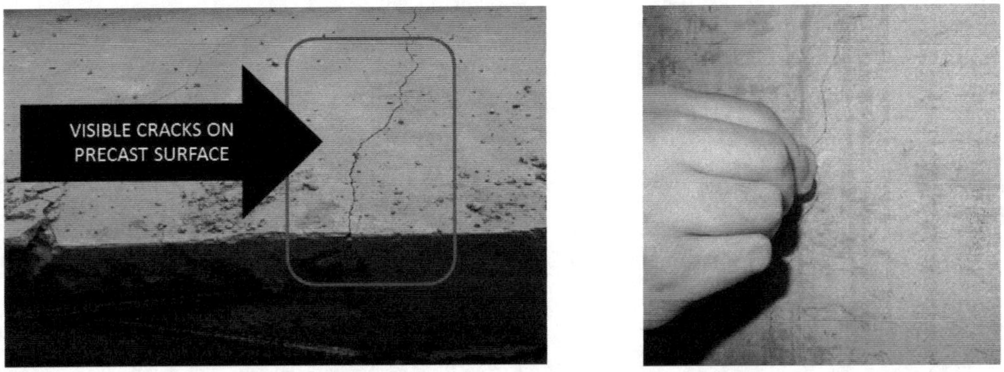

(a) (b)

Fig 1. (a) and (b) Visible cracks on Pre-cast surface.

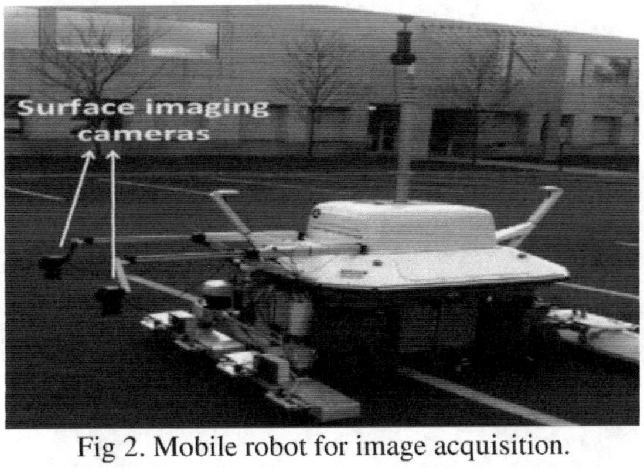

Fig 2. Mobile robot for image acquisition.

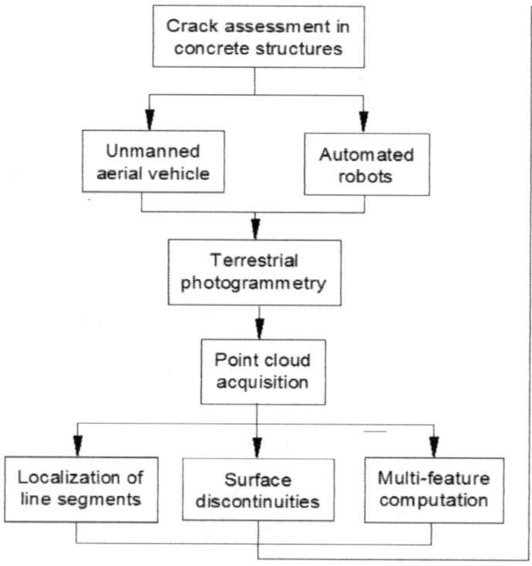

Fig 3. Flow chart to process crack detection and analysis in construction.

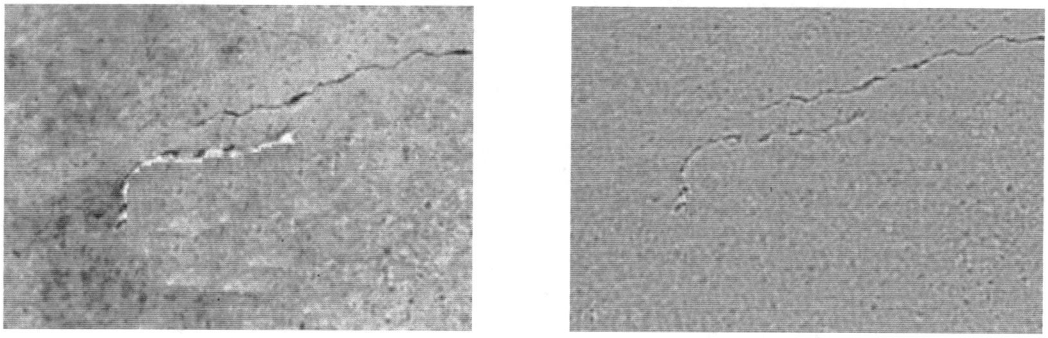

(a) (b)

Fig 4. (a) Original image with crack;(b) converted image with image processing for further analysis.

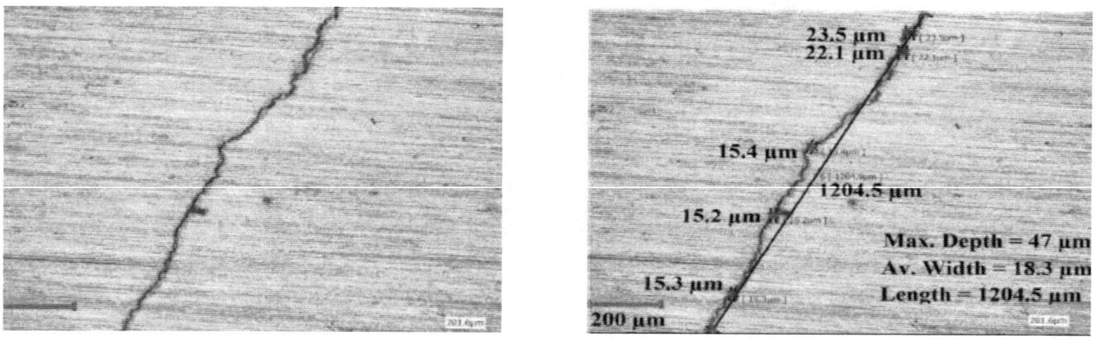

(a) (b)

Fig 5. (a) Acquired image through image processing;(b) density analysis of the formed crack [6].

IOP Conf. Series: Materials Science and Engineering **961** (2020) 012006 doi:10.1088/1757-899X/961/1/012006

3.2.Image processing in fluid flow applications

The exclusive pattern of the supersonic jet structure formation, propagating into a stationary gas at an ambient temperature, is shown in figure 6 [12]. The pattern is obtained from a supersonic jet, impacted on an ambient material and accelerated through a bow shaped obstacle, whereas, the outflow decelerates in a Mach disk or jet shock. Image processing is carried out with the radiation obtained from a point x-ray backlighting source, which develops a point-projection shadow of the developed experimental setup on an x-ray film. This type of technique only gives jet flow pattern images of short scale lengths. By using an electron beam, fluorescent images can be produced to characterize and analyze the propagation of hypersonic jets at a long scale [13]. Mach number, jet velocity and jet-to-ambient density ratio, was considered as the main output parameters to study the fluid flow pattern. To analyze the developed output parameters, an image based algorithm as shown in Figure 7 is adopted to indicate the way the curvature looks of the developed head [14].

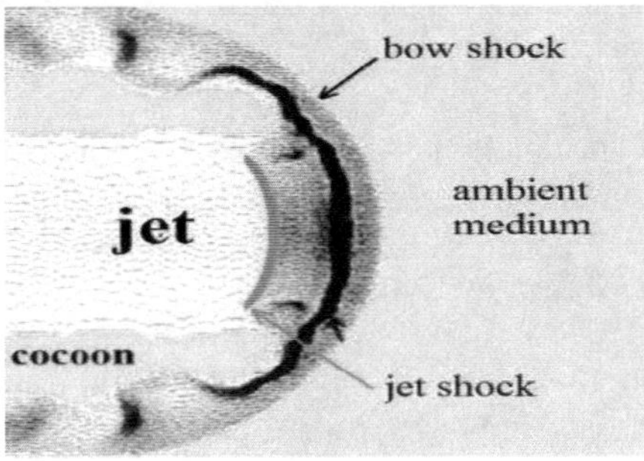

Fig 6. Structure of a jet head [14].

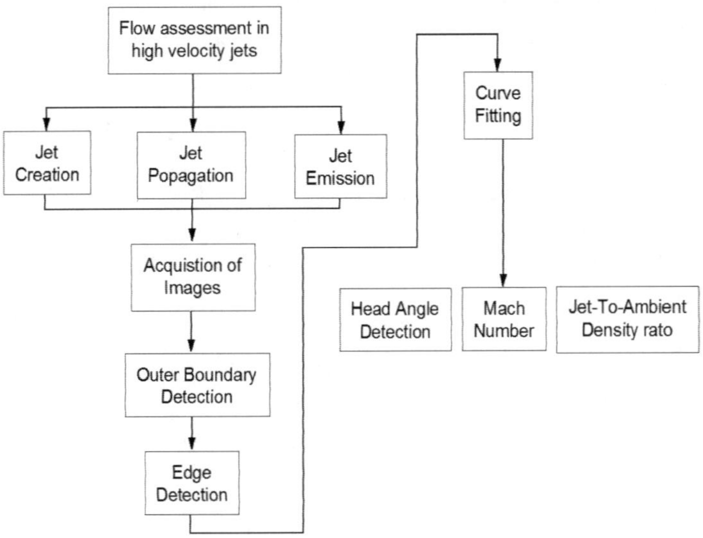

Fig 7. Algorithm used for image processing in fluid flow.

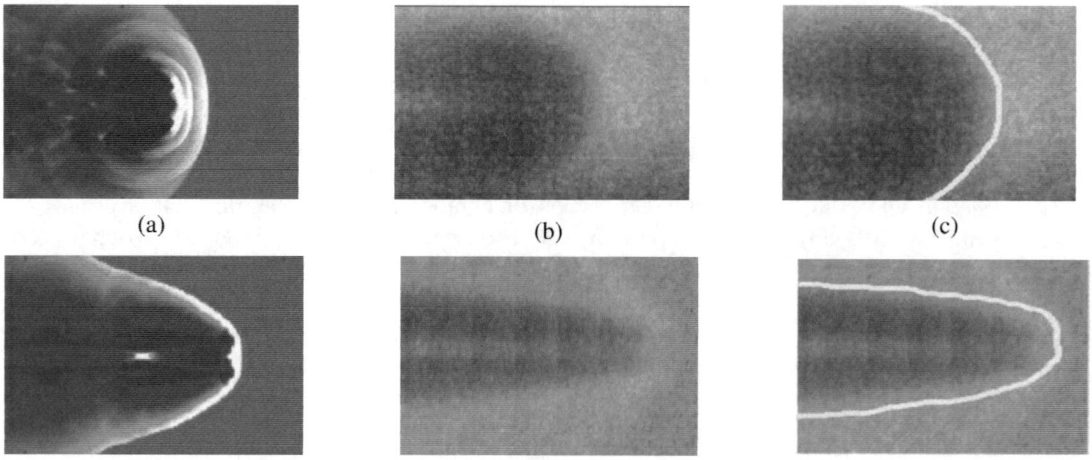

Fig 8. (a) Original image of jet initiation;(b) processed image of jet initiation;(c) image with curve fitting;(d) original image of jet propagation;(e) processed image of jet propagation;(f) image with curve fitting [15].

3.3.Image processing in thermal applications

A thermal imaging inspection in a refractory can even stop catastrophic failure leading loss in production and safety related problems. Thermal images can be also used for analyzing vehicle category in night time traffic. Thermal imaging cameras are most suitable in regions, where conventional cameras or image scanners could not be applied due to lack of illumination. Due to advancement in thermal imaging, application is also extended to border surveillance and security (cooled and uncooled cameras), considering their ability to detect sized targets in absolute darkness at extreme weather conditions. Application of thermal imaging are food industry, medicine, building diagnostics, tool condition monitoring, solar panels, volcanology, weather forecasting, Study of prototypes, by using thermal imaging cameras, may help the researchers and engineers to examine and determine the flaws in the prototypes and parts. Infrared technology can be applied for better and long lasting parts [16]. The flow chart given in Figure 9, clearly depicts the thermal imaging procedure that is mostly adopted.

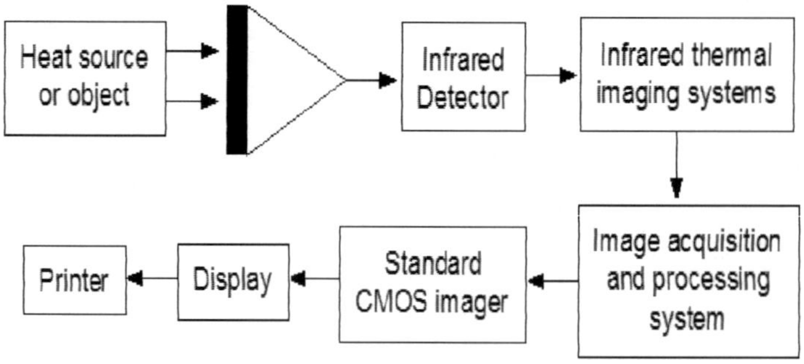

Fig 9. Steps of image processing used in thermal applications.

By the application of Kantorvich (S-K) algorithm, thermo-graphic images were reconstructed and their resolution has been enhanced.

3.4.Image processing in medical applications

Brain MRI can be used to diagnose glioma, HIV and cancer metastasis, in the similar way mammograms are used to detect breast cancer and CT scans are employed to detect cardiovascular diseases. Even the skin disorders like eczema, acne, melanoma, mycosis, etc., can also be recognized by microscopic images. The RGB scale of the images are taken into consideration to analyze the diseases. Different color shades using hue saturation value (HSV) and YCbCr like blackish, reddish, bluish, whitish and grayish are used to distinguish the region of interest. The contrast of the image can be adjusted more precisely by mapping the Region of Interests (ROI) with normalization of the gray image intensity. The whole MR and MRI images obtained can be stretched within the gray level region (0,255) and the noise in the images can be removed by normalizing the RGB within 0 to 1 using the following equations, where $g', R', G' and B'$ are the new values of gray level and RGB at the pixel i, j..

$$g'_{i,j} = \left[\left(g_{i,j} - mn \right) \frac{255}{mx - mn} \right] \quad (1)$$

$$R' = \frac{R}{R+G+B}, G' = \frac{G}{R+G+B}, B' = \frac{B}{R+G+B} \quad (2)$$

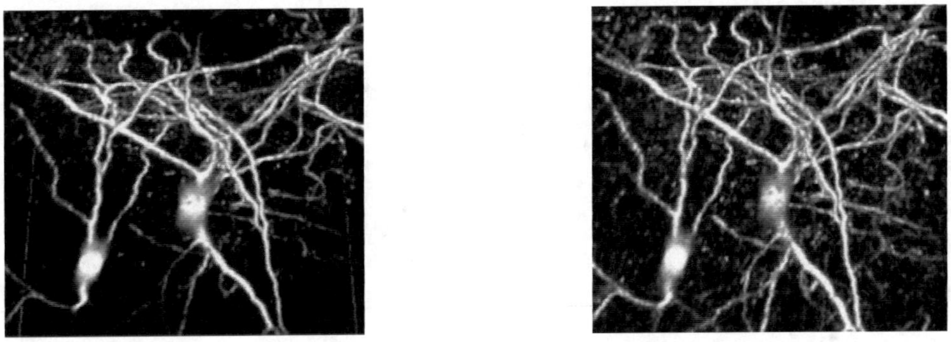

(a) (b)

Fig 10. (a) the original image of tissues;(b) the normalized image with full gray level (0,255)

This kind of process in medical field can be implemented even to study the skin disorders by normalizing the average skin color to gray intensity level so that the digitization errors can be minimized [17].

3.5.Image processing in material processing

From the previous studies of image processing in medical applications it is known that lot of re-constructional algorithms for 2D images into 3D images have been used [19,20]. Similar way, those relations are extended in the material processing. The relationship between the local 2D coordinate point (u,v) and the global 3D coordinate point P(x,y,z), using the camera coordinate system can be represented by

$$\lambda p = A[R][T]P \quad (3)$$

where λ is the scale parameter, p is the local coordinate point, A is the camera matrix which includes the focal length (fu,fv) and the mid-point of the image (u0,v0), and P is the global coordinate point. The R and T matrices are the rotational and transformation matrixes respectively.

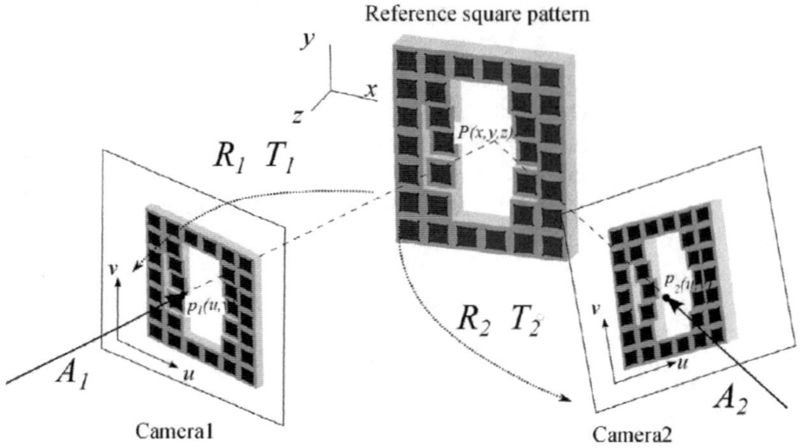

Fig 11. Comparison of rotation and translation of deformed pattern with reference pattern [18].

The A, R and T matrices are given by

$$A = \begin{bmatrix} f_u & s & u_o \\ 0 & f_v & v_0 \\ 0 & 0 & 1 \end{bmatrix}, R = \begin{bmatrix} r_{11} & r_{12} & r_{13} \\ r_{21} & r_{22} & r_{23} \\ r_{31} & r_{32} & r_{33} \end{bmatrix}, \text{and } T = \begin{bmatrix} T_x \\ T_y \\ T_z \end{bmatrix} \quad (4)$$

The part of A[R][T] can be condensed into a 3x4 matrix M.

$$M = \begin{bmatrix} M_{11} & M_{12} & M_{13} & M14 \\ M_{21} & M_{22} & M_{23} & M_{24} \\ M_{31} & M_{32} & M_{33} & M_{34} \end{bmatrix} \quad (5)$$

Once the rotation and the transformation of the etched pattern is known, then the position and the scaled image can be calculated, thus the strain developed on the sample can be obtained by the equation

$$\lambda \begin{bmatrix} u \\ v \\ 1 \end{bmatrix} = \begin{bmatrix} M_{11} & M_{12} & M_{13} & M14 \\ M_{21} & M_{22} & M_{23} & M_{24} \\ M_{31} & M_{32} & M_{33} & M_{34} \end{bmatrix} \begin{bmatrix} x \\ y \\ z \\ 1 \end{bmatrix} \quad (6)$$

Curve fitting method can be used to determine the dimensions of the developed strain on the sample where the patterns are made. By measuring the major and minor dimension of the deformed circle grids, and comparing with the original ones, the developed strain can be measured.

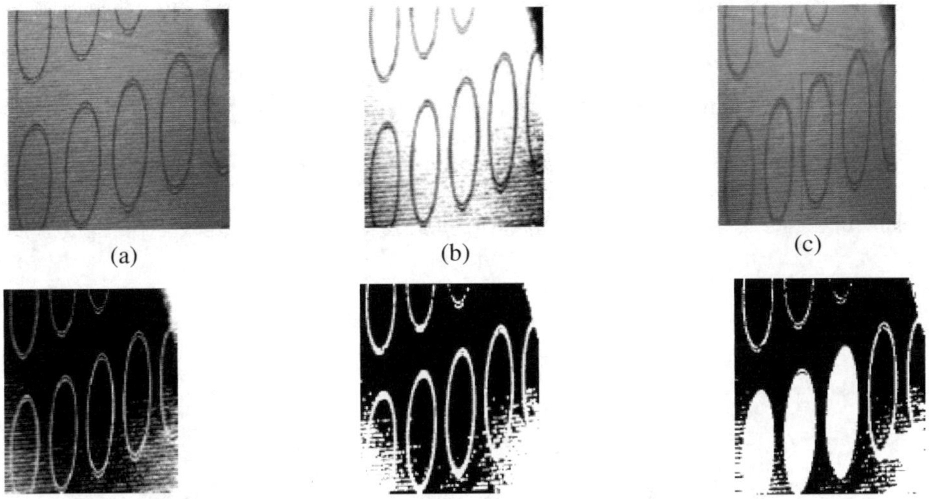

(a) (b) (c)

(d) (e) (f)

Fig 12. (a) Acquired image; (b) Gray scale to binary conversion; (c) Morphological element identification arbitrary shapes; (d) Complement of b/w image; (e) Two Dimensional Averaging filter and dilation and erosion to remove noises; (f) Flood fill.

To picking up the region of interest, the developed algorithm displays the major and minor axis dimensions. Using these dimensions, the developed strain can be calculated using the appropriate equations like,

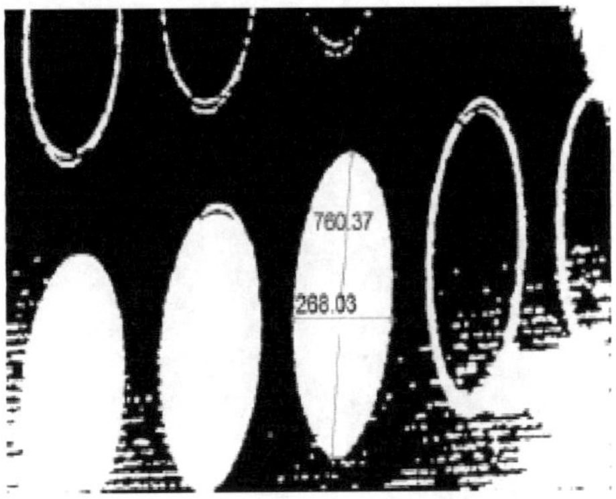

Fig 13. Measurement of image region major and minor axis dimension [22].

The major and minor strain is given by

$$\varepsilon_{major} = \ln\left(\frac{a}{d}\right) \text{ and } \varepsilon_{minor} = \ln\left(\frac{b}{d}\right) \quad (7)$$

where a and b denote the major and minor axis of the ellipse respectively, and d denotes the original diameter of the circle grid.

3.6.Image processing in food industries

Image processing is intended to use for food safety and standards. Moreover, this is a contemporary technique used to ensure consumer satisfaction mostly in a range of food-related fields. Mainly focusses on the detection of adulteration, ensures virgin fruits, vegetables and meat to customers. Even image processing is used to sever products and check & evaluate food-producing tools. This all eventually intense to meet quality in life-saving products, especially in foods.

Consider the fruit Orange which has infected regions in a different color in nature may be brown Figure 14 (a). By distinguishing these colors by Image processing the go or no go status of fruits will be decided. The pixels near the different colors or pixels which are a mixer of both (orange and brown) colors are identified and it is spotted and bounded by multicolor [23]. This multi-color represents the boundary to be detect called the region of interest (ROI).

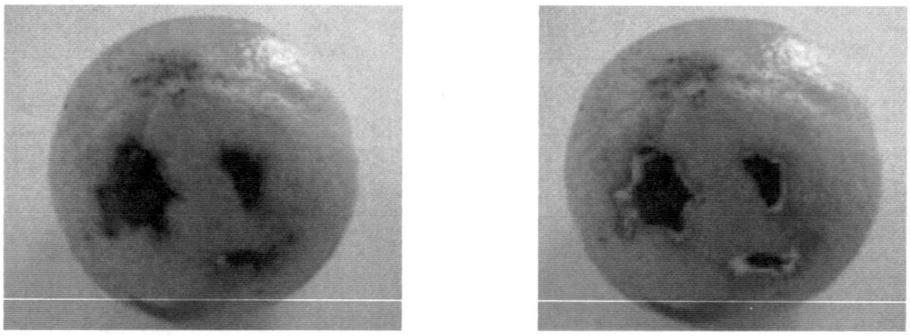

(a) (b)

Fig 14. Edge detection with gray intensity (a) level 100; (b) level 255.

Same as all other processes food industry also follows image capturing, image segmentation, object measurement or feature extraction, and classification. This method is pervasive to maintain a good quality of food. Fuzzy logic has also considered brightening the boundary. By adjusting the threshold value the RGB [25, 26] image can be converted to grayscale figure 15.

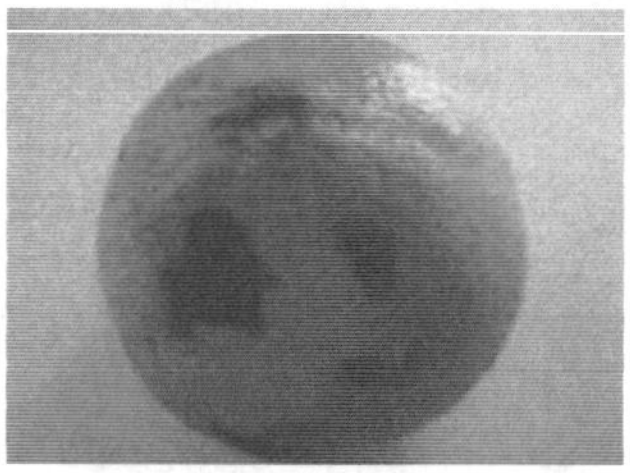

Fig 15. Normalized image with different thresholds of RGB.

IOP Conf. Series: Materials Science and Engineering **961** (2020) 012006 doi:10.1088/1757-899X/961/1/012006

Actually, the conversion of RGB to grayscale or black & white is to increase the processing time and reduce the computational time. Even color intensity will not have any influence over the results.

4. Conclusion

Computer aided Image processing techniques is pervasive in all the field and also a contemporary field of research. Several applications of image processing like construction, fluid flow, medical treatment, material screening, medical and food processing were studied in this review. Most of these applications adapt the same procedures like filtering the initial input image, segmentation and separation of the region of interest, area extraction and classification based on these required features. Although each area required special treatment, there are few common methods like edge detection, edge sharpening, noise detection and smoothening, and conversion to gray scale image for identification. All these are done through normalizing the RGB and gray values by adopting few algorithms like neural networks, fuzzy clustering, decision trees, random forests, etc. Eventually Potential for doing research in computer aided image processing techniques are substantial and challenges to be solved and yet to find in each field are boundless. The concept and methodology to implement these techniques is almost similar in all fields which can infer to the new researchers with wide opportunities. Eventually it is proven that substantial innovations and researches are in progress in and around the computer aided digital image processing and this going to lead the future unanimously.

References

[1] PrateekPrasanna, Kristin J Dana, NenadGuncunski, Basily B Basily and Hung M La 2014 Automated Crack Detection on concrete bridges, *IEEE Transactions on Automation Science and Engineering,* Vol. **13 (2)**, pp. 1-9.

[2] Valenca J, Puente I, Julio E, Gonzalez Jorge H and Arias Sanchez P 2017 Assessment of cracks on concrete bridges using image processing,*Laser scanning survey, Construction and Building Materials,* Vol. **146**, pp. 668-678.

[3] Jingsi Yang and Jesse Zhu 2015Cluster identification using image processing, *Particulogy*, Vol. **23**, pp. 16-24.

[4] Jin- Young Park, Jung-Ho Chung, Jung-Seok Lee, Hee-Jim Kim, Seong-Ho Chol and Ul-Won Jung 2017 Comparison of the diagnostics accuracies of optical coherence tomography, micro computed tomography, and histology in periodontal disease: an ex vivo study, *Journal of Periodontal and Implant Science*, Vol. **47(1)**, pp. 30-40.

[5] Jeong Choi, Yun Lee, Yun Yong Kim and Bang Yeon Lee 2017Image processing technique to detect carbonation regions of concrete sprayed with phenolphthalein solution, *Construction and Building Materials,* Vol. **145**, pp. 451-461.

[6] Hesham M Shehata, Yasser S Mohammed, Mohammed Abdellatif and Taher H Awad 2018 Depth estimation of steel cracks using laser and image processing techniques, *Alexandria Engineering Journal,* Vol. **57** (4), pp. 2713-2718.

[7] Maheswari Subramanian and ReebaKorah 2018 A framework of secured embedding scheme using vector discrete wavelet transformation and lagrange interpolation, *Journal of computer networks and communications,* Vol **20**, pp. 1-9.

[8] ChristoperSkea, AlirezaRezagholilou, Pouria Behnoud Far, RaoofGholami and Mohammed Sarmadivleh 2018An approach for wellbore failure analysis using rock carvings and image processing, *Journal of Rock Mechanics and Geotechnical Engineering,* Vol. **10 (5)**, pp. 865-878.

[9] Tomoyuki Yamaguchi, Shingo Nakamura, Ryo Saegusa and Shuji Hashimito 2008 Image based crack detection for real concrete surfaces, *IEE Journal on Transactions on Electrical and Electronics Engineering,* Vol. **3(1)**, pp. 128-135.

[10] Carnedente V 2013 Experimental and numerical simulation, by an arcjet facility, of hypersonic flow in Titan's atmosphere, *Experimental Thermal and Fluid Science,* Vol. **48**, pp. 97-101.

[11] KedarBharadwaj, AzadehSheidaei, ArashAfshar and JavadBaqersad 2019 Full field strain prediction using mode shapes measured with digital image correlation, *Measurement,*Vol. **139**, pp. 326-333.

[12] Blondin J M 1989 Herbig-Haro objects as the heads of radiative jets, *The Astrophysical Journal*, Vol. **37**, pp. 37-40.

[13] Belan M 2014 Hypersonic jets in astrophysical conditions: focus on spreading and asymmetric stability properties, *New Journal of Physics,* Vol. **16**, pp. 679-698.

[14] Mirazaei M, and Khodabakshi Rafsanjani H 2017 An automatic algorithm for determination of the nanoparticles from TEM images using circular hough transform, *Micron,* Vol. **96**, pp. 86-95.

[15] Mohsen Mirzaei, and Hossein Khodabakshi Rafsanjani 2018 Assessment of image processing methods to characterize the precise shape of the jet head in the hypersonic jet flows, *Flow Measurement and Instrumentation,* Vol. **64**, pp. 142-154.

[16] Xiaona Li, Ruolan Li, Mengyu Wang, Yaru Liu, Baohua Zhang and Jun Zhou 2017 Hyperspectral Imaging and Their Applications in the Nondestructive Quality Assessment of Fruits and Vegetables,*hyperspectral-imaging-in-agriculture-food-and-environment(Intech Open).*

[17] Soma Datta and NabenduChaki 2015 Person identification technique using RGB based on dental images, *14th Computer Information Systems and Industrial Management (CISIM)* Warsaw, Poland, pp.**169-180**,

[18] Kentaro Sato, Qiang Yu, Jiro Hiramoto, Toshiaki Urabe and Akihide Yoshitake 2015 A method to investigate strain rate effects on necking and fracture behaviors of advanced high strength steels using digital imaging strain analysis, *International Journal of Impact Engineering,* Vol. **75**, 2015, pp. 11-26.

[19] PadmajaGrandhe, Sreenivas Reddy Erra and VasumathiDevara 2018 Adaptive analysis and reconstruction of 3D DICOM images using enhancement based SBIR algorithm over MRI, *Biomedical Research,* Vol. **29(4)**, pp. 644-653.

[20] Sato J 2006 Recovering multiple view geometry from mutual projections of multiple cameras, *International Journal of Computer Vision,* Vol. **66**, pp. 123-140.

[21] Kenji Takada, Kentaro Sato and Ninshu Ma 2015 Fracture prediction for automotive bodies using a ductile fracture criterion and a strain dependent anisotropy model, *SAE International,* Vol **1**, pp. 1-10.

IOP Conf. Series: Materials Science and Engineering **961** (2020) 012006 doi:10.1088/1757-899X/961/1/012006

[22] Mugendiran V, AGnanavelbabu 2017 Comparison of plastic strains on AA5052 by single point incremental forming process using digital image processing, *Journal of Mechanical Science and Technology,* Vol. **31 (6)**, pp. 3943-2949.

[23] Nikos Petrellis 2018 A review of image processing techniques common in human and plant disease diagnosis, *Symmetry,* Vol. **10(7)**, pp. 1-34.

[24] Deepak P Hujare ,Nitesh R Girase and Madhuri G Karnik 2016 Crack detection in Shaft by Finite Element Analysis and Experimental Modal Analysis, *European Journal of Advances in Engineering and Technology,*Vol**3(10)**: 18-24.

[25] Muthukrishnan R 2011 Edge detection techniques for image segmentation, *International Journal of Computer Science and Information Technoology,* Vol. **3,** pp. 259-267.

[26] JissKuruvilla, Dhanya Sukumaran, Anjali Sankar and Siji P Joy 2016 A review on image processing and image segmentation, *International Conference on Data Mining and Advanced Computing (SAPIENCE).*

Effect of Microsilica on Strength and Microstructure of the GGBS-based Cement composites

M N Khan[1], S Singla[1*], R Garg[2], and R Garg[3]

[1]Department of Civil Engineering, RIMT University, Mandi Gobindgarh, India
[2]Department of Civil Engineering, Galgotias College of Engineering and Technology, Greater Noida, India
[3]Department of Chemistry, Rayat Bahra University, Mohali, India
*Corresponding author: drsandeepsinglaz@gmail.com

Abstract. Cement production requires substantial energy and significantly accounts for global carbon dioxide emissions. Thus, the consumption of ordinary portland cement (OPC) must be reduced through incorporation of auxiliary materials. Also, the mechanical strength and durability of structures, in the construction industry, need to be improved for economizing the maintenance cost and increasing the service life. This study explores the effect of partial substitution of cement by granulated blast furnace slag (GGBS), and microsilica (MS), the industrial by-products. This study focuses on the use of 10% GGBS and 0-16% MS as a substituent of cement at a water binder ratio of 0.42. The fresh properties were determined to study the effect of these substituents. The compressive strength of all the mixes was determined after 3, 7, 28, and 56 days of treatment. The results were correlated through microstructural analysis. The study revealed that the cement composites with an optimal substituent dosage of 10% GGBS and 12% MS can attain adequate compressive strength and can be used for practical applications.

1. Introduction

Construction technology can aid in sustainable development by using less energy and natural resources along with the production of lesser waste material. The concern also focuses on the improvement of fresh as well as hardened properties of the mortar. The quality of cement mortar is characterized by its mobility, retention of water, strength, and durability and depends on its ingredients [1]. A good mortar should be economical, easy to work, well-adherent, durable, and chemical attack resistant. As per ideal requirements, the paste should not only be workable, but the constructed material should be durable with high mechanical strength. Particle size distribution (PSD) controls the flow of mortar, while the mechanical properties and durability are controlled by the mix grading and the resultant packing in the matrix. Further, the hydration behavior and the requirement for flow is dependent upon the extent of cement dispersion in water[2]. The flocculation level of a cement paste is often related to water content and early hydration. PSD, matrix packing, and water demand in cement mortar can be controlled by using admixtures such as water-reducing compounds and mineral admixtures [3].

Sustainable development is described as the development that meets the needs of the present generation without sacrificing next-generation needs. The production of the prime constituents of cement mortar i.e. cement and fine aggregates pose an environmental threat. Likewise, the use of an enormous amount of drinking water leads to scarcity of these resources in the long run [4]. As per estimation, one ton of cement clinker production yields one ton of greenhouse gases (GHGs) such as carbon dioxide responsible for global warming. The production of fine aggregates requires extraction from quarries at the expenditure of energy with the emission of waste products leading to environmental pollution. Thus, the construction industry is not only one of the largest consumers of energy and natural resources but also an immediate concern for environmental issues [5]. By incorporating different mineral admixtures such as metakaolin, GGBS, fly ash, rice husk powder, palms oils, and silica fumes, concrete consistency can be improved. The hardness of concrete is also influenced by mineral admixtures. The integration of these admixtures reduces cement content which reduces the environmental impact and also improves the properties of concrete [6]. Waste disposal problems can also be reduced as they are industrial by-products. An improvement in the mechanical properties and toughness of cement indicates the positive influence of mineral mixtures on concrete [7]. The use of these additional cementing materials in concrete preparation will lead to substantial improvements in resources, cost efficiency, and a decrease in environmental emissions [8].

Microsilica (MS) has been found to enhance the physical and mechanical aspects of cement mortar to a great extent [9]. MS is an important pozzolanic and cementitious material consisting of very fine and non-crystalline silica obtained by the processing of silicon- containing alloys or elemental silicone in electric arc furnaces [10]. MS particles have an average microsphere diameter, of roughly 0.15 microns, and primarily belongs to the cristobalite, vitreous form of silica. On average, one microsphere weighs 100 times lesser than a typical grain of cement. Owing to its fine quality and rich content of silicon dioxide, it plays an important role in pozzolanic action in the cement matrix (Equation 1 and 2).

$$2SiO_2 + 3Ca(OH)_2 \rightarrow \underset{CSH}{3CaO.2SiO_2.3H_2O} \quad \text{(Equation 1)}$$

$$2SiO_2 + 3CaO + 3H_2O \rightarrow 3CaO.2SiO_2.3H_2O \quad \text{(Equation 2)}$$

Microsilica results in enhancement of CSH gel production and refined pore structure, thereby reducing permeability and improving the durability of cement mortar [11]. It also results in a hindrance to the diffusion of detrimental ions and increases the resistance towards an aggressive environment [12].

Ground granular blast furnace slag (GGBS) is a mineral admixture, generally extracted from molten slag as a granular substance that is grounded after trying to get a powdery form. GGBS has pozzolanic and cementitious properties [13]. GGBS has unique characteristics in terms of resistivity to corrosion, attack of sulfate, and permeability of water. GGBS reduces creep and bleeding but increases the heat of hydration, resulting in improved compressive strength of the mortar. GGBS is an aluminosilicate with main constituents as SiO_2, Al_2O_3, CaO, and MgO [14]. The addition of GGBS was found to improve the workability and the long-term compressive strength [15]. The incorporation of SF and granulated blast furnace slag was found to provide lower early age strength to coral aggregate-concrete but higher late age strength due to late initiation of the pozzolanic reaction [16]. Auxiliary addition of GGBS and SF was found to boost the mechanical characteristics of fly ash-based concrete [17]. The combination (MS 10%+ GGBS 30%) was found to provide a more workable and higher strength relative to standard concrete for all curing days [18]. The replacement of cement by SF and GGBS in presence of fly ash and metakaolin was found to increase the consistency and retard the setting time [19]. The optimal dosage of SF and GGBS has been obtained as 10% for the best workability and strength of concrete [20]. The incorporation of 10% SF and 10% GGBS was found to provide maximum compressive strength and abrasion resistance of concrete [21]. Self-compacting concrete was found to be more cost-effective and with better strength with the incorporation of GGBS, SF, and metakaolin [22]. The presence of auxiliary GGBS and SF was also found to improve the compressive strength of otherwise weak recycled aggregate- based concrete [4].

In this study, the effect of partial substitution of cement with MS in presence of GGBS was analyzed on the fresh properties, strength, and microstructure of cement mortars at primitive and advanced ages. GGBS was incorporated at 10% by weight of cement as a partial substituent of cement. MS content was varied as 4%, 8%, 12%, and 16% by weight for partial substitution of cement. The microstructural analysis of cement mortars with and without MS was carried out using Scanning Electron Microscope (SEM) studies.

2. Materials and Methods

The cement used for the experimental studies was ordinary Portland cement (43 Grade). GGBS and Micro-silica were obtained as a by-product of the local industry and was used as such without any modification. The Chemical composition of the binder materials have been given in Table 1. Fine aggregates were procured from the locally available source and conformed to IS: 383-1970 [18]. The physical characteristics of components of mortar have been given in Table 2. 10% GGBS was used as a fixed substituent by weight of cement for all the mixes, except the control mix (CM). Besides, cement was substituted partially by 4%, 8%, 12% and 16%MS by weight of cement as per mix composition given in Table 3. The consistency and setting time of the cement was determined through Vicat's Apparatus and flow table was used for flow determination of mortars [19]. The mortar was prepared with a binder: sand ratio of 1:3 with a water-binder ratio at 0.42. The components were mechanically mixed and moulded into cubes with proper compaction to eliminate furthermore air voids. The mortar cubes were allowed to set for 24 hours and then placed in a water tank for curing at standard temperature. Compressive strength is a key value for construction practices. The compressive strength of the fully cured, surface dried cement mortar specimens was measured with a compression testing machine at 3 days, 7days, 28 days, and 56 days. Testing was carried out with three cubes and the average value was considered [18]. Microstructural analysis was carried out through Philips XL20 Scanning Electron Microscope.

Table 1. Chemical Composition of Binder Materials

Oxides % by mass	OPC % by mass	MS % by mass	GGBS % by mass
SiO_2	19.46	98	35.6
Al_2O_3	4.22	0.03	13.19
Fe_2O_3	3.56	0.02	1.13
CaO	65.92	0.3	41.94
MgO	1.08	0.2	3.76
K_2O	0.67	0.3	0.41
Na_2O	0.21	0.3	0.32
TiO_2	0.24	-	0.52
SO_3	2.6	-	1.5
Cl	0.01	-	0.01
LOI	0.96	0.8	1.41

Table 2. Physical Properties of Components

Aggregate Type	Bulk Density (kg/m^2)	Fineness (m^2/kg)	Specific gravity
GGBS	1000-1100	350-550	2.90
MS	1420-1540	15000-35000	2.22
Fine aggregate	1615	1.15	2.55

Table 3. Mix proportion of Specimens per m^3

Designation	Cement (g)	Sand (g)	GGBS (g)	MS (g)	Water (g)
CM	200	600	-	-	84
MS1	172	600	20	8	84
MS2	164	600	20	16	84
MS3	156	600	20	24	84
MS4	148	600	20	32	84

3. Results and Discussions

3.1. Fresh Properties

The observed values of consistency of mortars, without and with the replacement of cement with GGBS and MS have been plotted in figure 1. The consistency of mortars increased with the incorporation of GGBS and with an increasing percentage of MS from 4% to 16%. The consistency of the mortar is related to the size and specific surface of the particles present in the mortar [8]. The microparticles of GGBS and MS have a greater specific surface of the particles. As the percentage of MS increases, the hydration process is faster and the water required for the hydration increases, increasing the consistency of mortars [19].

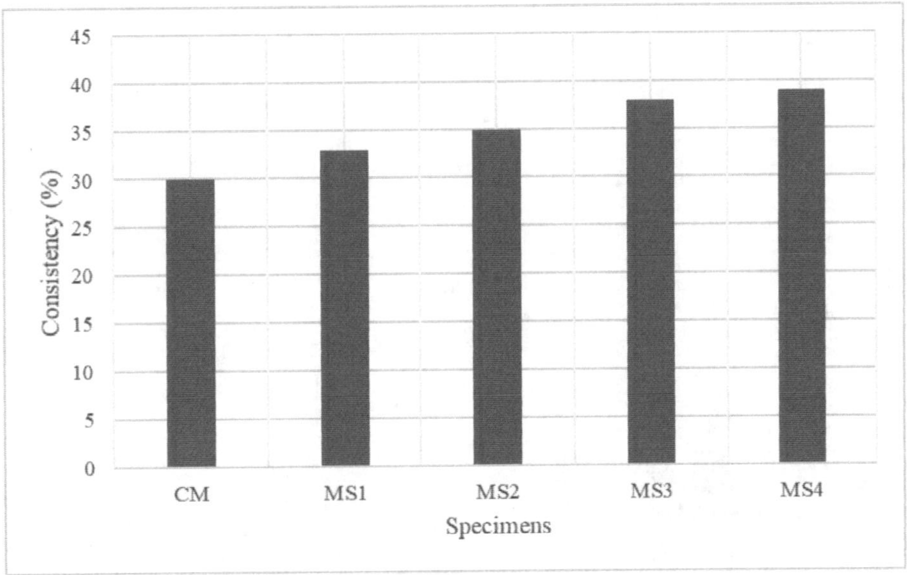

Figure 1. Consistency of Mortar Specimens

Figure 2 illustrates the effect of GGBS and MS on the setting time of mortars. Initial setting time (IST) as well as final setting time (FST) of mortars, increased with substitution of cement with GGBS, and with an increasing percentage of MS from 4% to 16%. The retardation effect has been attributed to the decreasing content of cement with the auxiliary GGBS and MS. The setting time is related to the extent of hydration reaction in the mortar. The literature reports the delay of hydration reaction by GGBS [19]. The fine particles of MS fill the voids in the matrix, hence, the porosity of the mortars decreased and the setting process gets decelerated, resulting in an increase in setting time of mortars [23]. The observed values of the flow of mortars, without and with the replacement of cement with GGBS and MS have been plotted in figure 3. The flow of mortars was found to decrease with auxiliary GGBS and with an increasing percentage of MS from 4% to 16%. The flow of the mortar is also related to the size and specific surface of the binder particles [24]. The fine particles of GGBS and

MS have a greater specific surface of the particles and result in better particle size distribution in the mixture. The filler effect of these particles reduces the pores and improve the packing of constituents in the matrix. As the percentage of MS increases, cohesive binding results in enhanced packing and the empty spaces in the matrix are better filled. As a result, water cannot bleed over the surface of the matrix. This increases the stiffness of the mortars resulting in a decrease in the flow of mortars. Similar effects have been reported in the literature [20].

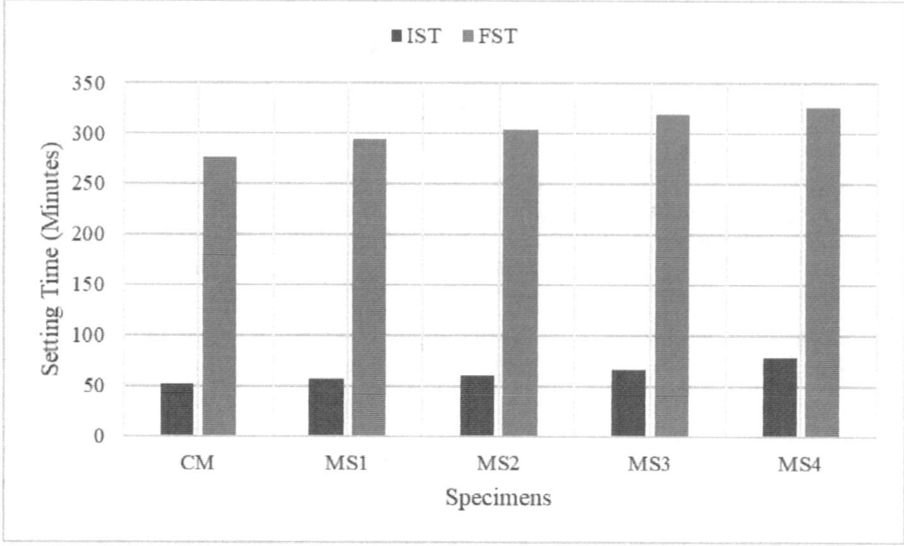

Figure 2. Setting Time of Mortar Specimens

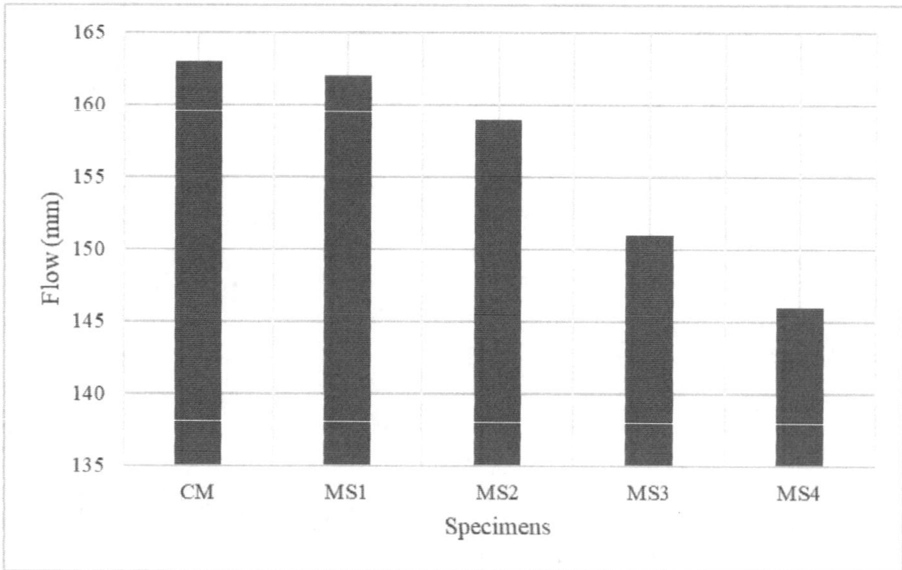

Figure 3. Flow of Mortar Specimens

3.2. Compression Strength

The effect of partial substitution of cement by GGBS and MS on the compressive strength of mortar at the age of 3, 7, 28, and 56 days has been analyzed in comparison to CM and represented in figure 4. The compressive power of all mortar mixtures, MS1, MS2, MS3, and MS4 at 3 days were higher compared to CM. Nevertheless, it has been observed that the compressive power of mortars has improved with substitution by GGBS as well as with an increasing percentage of MS from 4% to 12 % but decreased at 16% percentage of MS.

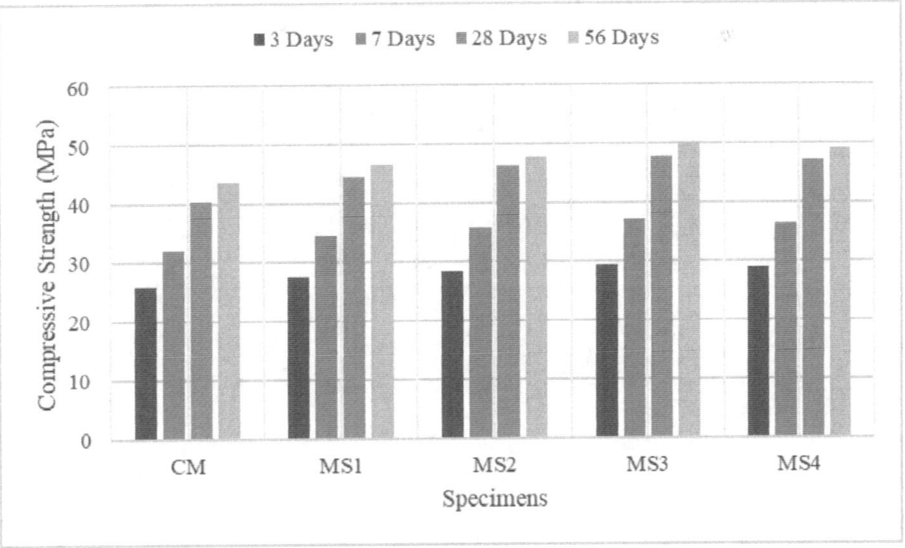

Figure 4. Compressive Strength of Mortar Specimens

The compressive strength of the CM at the age of 3 days was 25.96 MPa. The mix MS1, MS2, and MS3 showed an increase of 5.97%, 9.17%, and 13.33%, but MS4 showed an increase of 12.10% as compared to CM. For MS3, the compressive strength was obtained as 29.42 MPa, maximum among all mixes. At 7 days of age, the compressive strength of the CM was 32 MPa. The mix MS1, MS2, and MS3 showed an increase of 7.72%, 11.56%, and 16.22%, but MS4 showed an increase of 13.53% as compared to CM. For MS3, the compressive force was measured as 37.19 MPa. At 28 days, the compressive strength of CM was 40.37 MPa. The mix MS1, MS2 and MS3 showed an increase of 10.45%, 15.06%, and 18.68%, but MS4 showed an increase of 17.17% as compared to CM. In this case, the maximum compressive strength was observed for MS3 as 47.91 MPa. At 56 days of curing age, the compressive strength of CM was 43.76 MPa. The mix MS1, MS2, and MS3 showed an increase of 6.44%, 9.28%, and 15.04%, but MS4 showed an increase of 12.68% as compared to CM. The maximum compressive strength was observed for MS3 as 50.34 MPa. The comparative study reveals that the mortars mixes, with the replacement of cement by 10% GGBS and 12% MS exhibited increased compressive strength than the controlled mix, CM. This observation can be attributed to the pozzolanic activity of MS and GGBS [21]. Cement hydration creates calcium hydroxide and contributes to the formation of pores in the cement matrix. Silica particles in MS and GGBS react with calcium hydroxide to form calcium silicate hydrate gel that gets filled in the pores of the cement matrix. This increases the compressive strength of the cement mortar [10]. Besides, as the percentage of MS is increased in the cement matrix, it increases the pozzolanic effect of MS and improves its compressive strength. The sudden drop in compressive strength at a higher percentage of MS may be due to the reason that a higher percentage of MS can result in friction among particles leading to reduced packing and hence reduced compressibility [25]. Thus, 12% MS in presence of 10% GGBS is the optimal substituent dosage for the studied systems. The data was further comparatively analyzed in terms of age of curing days. An increase in compressive strength was observed in all the mortar mixes as evident in figure 5. The comparative account for percentage increase in compressive strength has been listed in Table 4. The data revealed that as compared to CM, the percentage increase of compressive strength in mortars with the replacement of cement with GGBS and MS was higher at increasing days of curing age upto 28 days. This observation confirms that the pozzolanic activity of

GGBS and MS strengthens the mortar mixes and hence increases the strength of such mixes as compared to CM. Similar trends have been reported by other researchers during their investigations [16,26].

Table 4. % Increase in Compressive Strength of Specimens

Specimen	3 Days	7 Days	28 days	56 Days
MS1	5.97	7.72	10.45	6.44
MS2	9.17	11.56	15.06	9.28
MS3	13.33	16.22	18.68	15.04
MS4	12.10	13.53	17.17	12.68

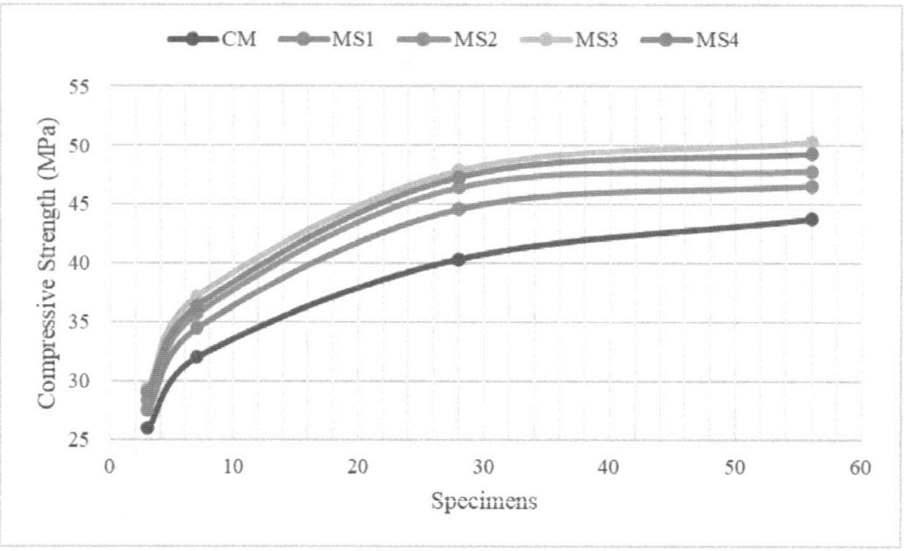

Figure 5. Effect of Curing age on Compressive Strength of Mortar Specimens

3.3. SEM Analysis

The SEM analysis of all the mortar samples including CM, MS1, MS2, MS3, and MS4 were carried out for a comparative account of the difference in microstructure with partial substitution of cement by GGBS and MS. The microstructure of CM at 3 days of curing age exhibits that the structure is porous and many $Ca(OH)_2$ crystals are present (Figure 7a).

Figure 6. Micrographs of CM at (a) 3 Days; (b) 7 Days; (c) 28 Days

The structure is still porous after 7 days of curing age, with some cracks but comparatively lesser $Ca(OH)_2$ crystals are present (Figure 7b). Even after 28 days of curing, the cement matrix appears porous accounting for the low compressive strength (Figure 7c). The micrograph of MS1 at 3 days of curing shows some pores in the structure and calcium hydroxide crystals. Thus, the microstructure of MS1 with cement replaced by 4% MS in presence of 10% GGBS appears loose and that the calcium hydroxide has not been consumed much at the age of 3 days (Figure 8a). It shows that the pozzolanic reaction is still delayed which accounts for the only slight increase in strength as compared to that in the structure of CM [27]. The microstructure after 7 days shows comparatively fewer pores and also shows the calcium silicate hydrate crystals due to the commencement of the pozzolanic reaction (Figure 8b). However, the microstructure after 28 days appears comparatively denser and shows many calcium silicate hydrate crystals leading to high compressive strength as compared to CM (Figure 8c).

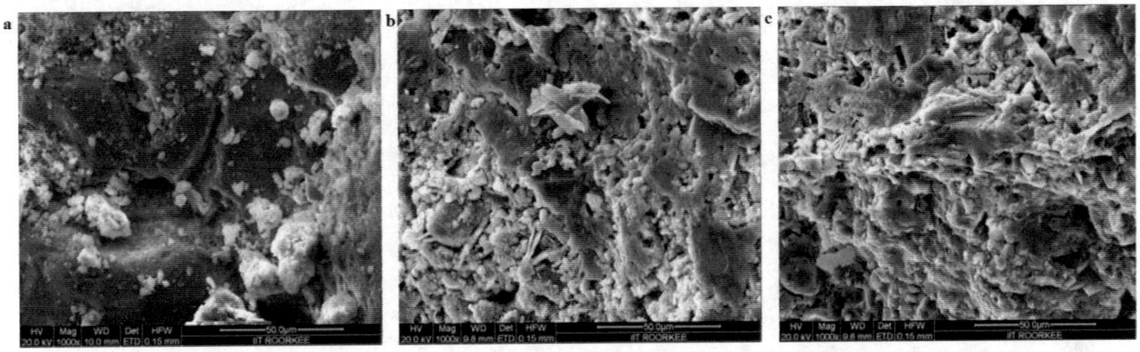

Figure 7. Micrographs of MS1 at (a) 3 Days; (b) 7 Days; (c) 28 Days

The microstructure of MS2 with cement replaced by 8% MS in presence of 10% GGBS is compact with lesser pores and that the calcium hydroxide has been quite consumed at the age of 3 days showing the initiation of the pozzolanic reaction accounting for the increased strength as compared to that in CM and MS1 (Figure 9a). With progression in curing age, the structure appears comparatively denser and shows many calcium silicate hydrate crystals leading to high compressive strength (Figure 9b and 9c).

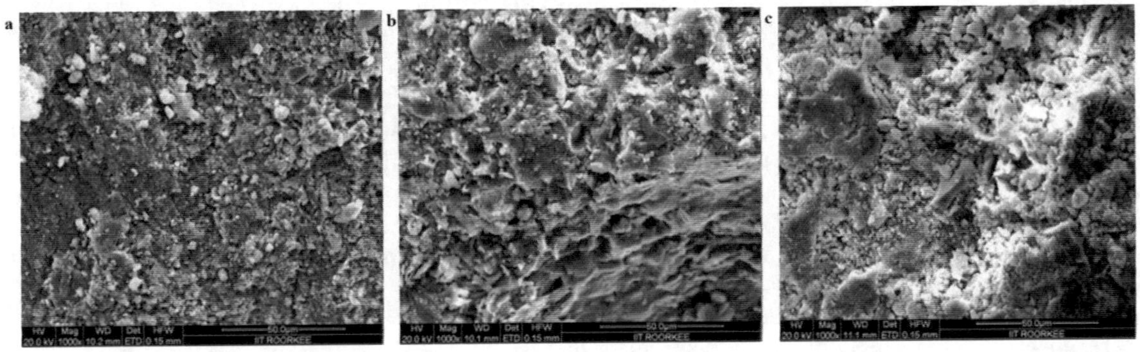

Figure 8. Micrographs of MS2 at (a) 3 Days; (b) 7 Days; (c) 28 Days

The micrograph of MS3 (12% MS in presence of 10% GGBS) shows scanty pores in the structure and calcium hydroxide crystals as compared to that in the structure of CM, MS1, and MS2 (Figure 10). Thus, the microstructure of MS3 with cement replaced by 12% MS in presence of 10% GGBS is quite compact and that the calcium hydroxide has been quite consumed with progressing curing age showing the development of the pozzolanic reaction accounting for the increased strength [10]. The microstructures of MS4 are comparatively less homogeneous at the studied ages consistent with the strength analysis (Figure 11).

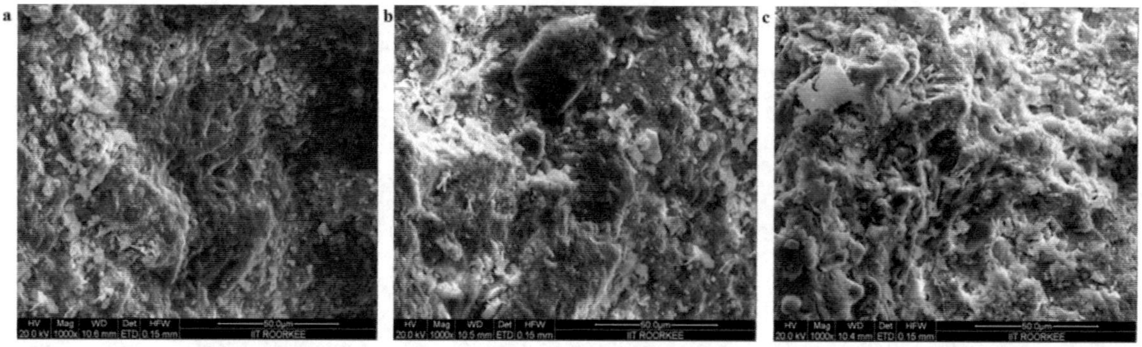

Figure 9. Micrographs of MS3 at (a) 3 Days; (b) 7 Days; (c) 28 Days

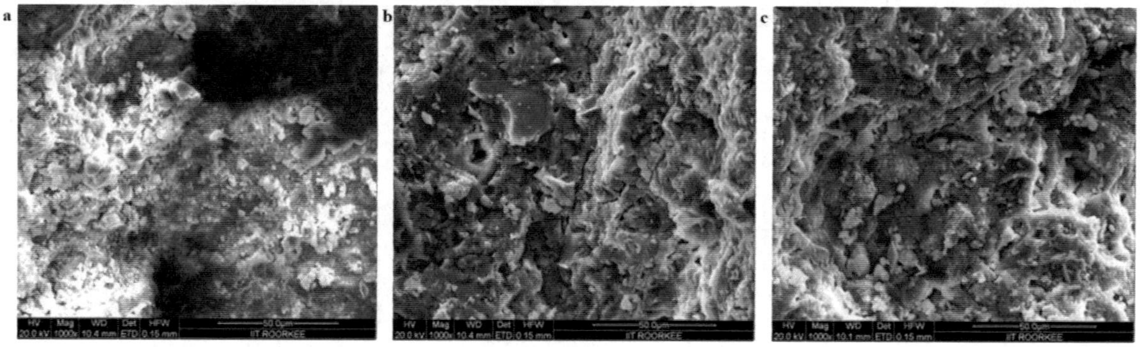

Figure 10. Micrographs of MS4 at (a) 3 Days; (b) 7 Days; (c) 28 Days

4. Conclusion

Based on the study, the following conclusions can be drawn: 1. Replacement of cement with GGBS and MS in presence of GGBS leads to an increase in consistency and delay of the setting process. This also increases the stiffness of the mortar and decrease of the flow. 2. The mixes with the replacement of cement by MS in presence of 10% GGBS showed an increasing order of compressive strength with maximum value at 12% MS. 3. The microstructure of mortar mixes without replacement of cement with MS and GGBS appeared more porous and less compact leading to lesser compressive strength. On contrary, the microstructure of mortar with 12% MS in presence of 10% GGBS appeared least porous and most compact with calcium silicate hydrate gel leading to higher compressive strength. Further, the combined cost of optimized content of GGBS and MS is comparable to that of cement and gives commendable strength enhancement. Thus, the partial substitution of cement by optimal amount of MS (12%) in presence of GGBS (10%) can lead to enhancement of compressive strength along with reduced cement use leading to lesser environmental hazards.

References

[1] Seleem H E D H, Rashad A M and El-Sabbagh B A 2010 Durability and strength evaluation of high-performance concrete in marine structures *Constr. Build. Mater.* **24** 878–84
[2] Osman D A M, Nur O and Mustafa M A 2020 Reduction of Energy Consumption in Cement Industry Using Zinc Oxide Nanoparticles *J. Mater. Civ. Eng.* **32** 1–13
[3] Kırgız M S 2016 Fresh and hardened properties of green binder concrete containing marble powder and brick powder *Eur. J. Environ. Civ. Eng.* **20** s64–101

[4] Guo Z, Jiang T, Zhang J, Kong X, Chen C and Lehman D E 2020 Mechanical and durability properties of sustainable self-compacting concrete with recycled concrete aggregate and fly ash, slag and silica fume *Constr. Build. Mater.* **231** 117115

[5] Mohamed O, Khattab R and Alzo'Ubi A K 2019 Factors Affecting Compressive Strength Development in Alkali-activated Slag Concrete *IOP Conf. Ser. Mater. Sci. Eng.* **603**

[6] Kanamarlapudi L, Jonalagadda K B, Jagarapu D C K and Eluru A 2020 Different mineral admixtures in concrete: a review *SN Appl. Sci.* **2** 1–10

[7] Mohamed O A 2019 Effect of mix constituents and curing conditions on compressive strength of sustainable self-consolidating concrete *Sustain.* **11**

[8] Heikal M, El-Didamony H, Sokkary T M and Ahmed I A 2013 Behavior of composite cement pastes containing microsilica and fly ash at elevated temperature *Constr. Build. Mater.* **38** 1180–90

[9] Yongo E, Manyala J ., Kito K, Matsushita Y, Outa N . and J.M N 2016 Diet of Silver Cyprinid, Rastrineobola argentea in Lake Victoria, Kenya *Int. J. Adv. Res.* **4** 144–9

[10] Dadsetan S and Bai J 2017 Mechanical and microstructural properties of self-compacting concrete blended with metakaolin, ground granulated blast-furnace slag and fly ash *Constr. Build. Mater.* **146** 658–67

[11] Rishav Garg, Manjeet Bansal and Yogesh Aggarwal 2016 Split Tensile Strength of Cement Mortar Incorporating Micro and Nano Silica at Early Ages *Int. J. Eng. Res.* **V5**

[12] Ganjian E and Pouya H S 2005 Effect of magnesium and sulfate ions on durability of silica fume blended mixes exposed to the seawater tidal zone *Cem. Concr. Res.* **35** 1332–43

[13] Ganesh Babu K and Sree Rama Kumar V 2000 Efficiency of GGBS in concrete *Cem. Concr. Res.* **30** 1031–6

[14] Pal S C, Mukherjee A and Pathak S R 2003 Investigation of hydraulic activity of ground granulated blast furnace slag in concrete *Cem. Concr. Res.* **33** 1481–6

[15] Lee Y, Lim S and Lee H 2020 Chloride Resistance of Portland Cement-Based Mortar Incorporating High Aluminate Cement and Calcium Carbonate *Materials (Basel).* **13** 359

[16] Wang Y, Zhang S, Niu D, Su L and Luo D 2019 Effects of silica fume and blast furnace slag on the mechanical properties and chloride ion distribution of coral aggregate concrete *Constr. Build. Mater.* **214** 648–58

[17] Liu Y, Zhou X, Lv C, Yang Y and Liu T 2018 Use of Silica Fume and GGBS to Improve Frost Resistance of ECC with High-Volume Fly Ash *Adv. Civ. Eng.* **2018**

[18] Vijaya Bhaskar Reddy S and Srinivasa Rao D P 2016 Experimental studies on compressive strength of ternary blended concretes at different levels of micro silica and ggbs *Mater. Today Proc.* **3** 3752–60

[19] Dave N, Misra A K, Srivastava A and Kaushik S K 2017 Setting time and standard consistency of quaternary binders: The influence of cementitious material addition and mixing *Int. J. Sustain. Built Environ.* **6** 30–6

[20] Mohan A and Mini K M 2018 Strength Studies of SCC Incorporating Silica Fume and Ultra Fine GGBS *Mater. Today Proc.* **5** 23752–8

[21] Rashad A M, Seleem H E-D H and Shaheen A F 2014 Effect of Silica Fume and Slag on Compressive Strength and Abrasion Resistance of HVFA Concrete *Int. J. Concr. Struct. Mater.* **8** 69–81

[22] Vivek S S and Dhinakaran G 2017 Fresh and hardened properties of binary blend high strength self compacting concrete *Eng. Sci. Technol. an Int. J.* **20** 1173–9

[23] Rao G A 2003 Investigations on the performance of silica fume-incorporated cement pastes and mortars *Cem. Concr. Res.* **33** 1765–70

[24] Senff L, Hotza D, Repette W L, Ferreira V M and Labrincha J a 2010 Effect of nanosilica and microsilica on microstructure and hardened properties of cement pastes and mortars *Adv. Appl. Ceram.* **109** 104–10

[25] Hu C, Han Y, Gao Y, Zhang Y and Li Z 2014 Property investigation of calcium-silicate-

hydrate (C-S-H) gel in cementitious composites *Mater. Charact.* **95** 129–39
[26] Jo B W, Chakraborty S, Lee S-T and Lee Y S 2019 Durability Study of Silica Fume-mortar exposed to the Combined Sulfate and Chloride-rich Solution *KSCE J. Civ. Eng.* **23** 356–66
[27] Pelisser F, Zavarise N, Longo T A and Bernardin A M 2011 Concrete made with recycled tire rubber: Effect of alkaline activation and silica fume addition *J. Clean. Prod.* **19** 757–63

Investigation on Mechanical Strength of Cellular Concrete in Presence of Silica Fume

G M Fani[1], S Singla[1*], R Garg[2], and R Garg[3]

[1]Department of Civil Engineering, RIMT University, Mandi Gobindgarh, India
[2]Department of Civil Engineering, Galgotias College of Engineering and Technology, Greater Noida, India
[3]Department of Chemistry, Rayat Bahra University, Mohali, India
*Corresponding author: drsandeepsinglaz@gmail.com

Abstract. Over the past decade, Cellular Lightweight concrete has gained popularity in the construction industry. Despite its advantageous low weight, thermal insulating, and fire-resistant nature, its comparatively low strength is a matter of concern. To develop the strength of this type of concrete, a thorough characterization of its main characteristics must be studied thereafter in addition to supportive cementitious materials. The effect of polyurethane on density and strength of cellular concrete subjected to varying curing conditions have been investigated in this report. Besides, Silica Fume has been used as a partial substituent of cement and as supportive cementitious material. The cellular concrete has been obtained by varying silica fume (0-20% by weight of cement) and polyurethane (0-20% by volume) at 7, 28, 56, and 90 days of curing. Density, compressive strength, split tensile strength and flexural strength have been investigated for all the specimens. The findings revealed a decrease in density and strength of cellular concrete in the absence of SF and the presence of polyurethane. The introduction of silica fume resulted in higher density and strength revealing the beneficial use of this pozzolanic material.
Keywords: Cellular concrete, Silica fume, Polyurethane, Compressive Strength, Split tensile strength.

1. Introduction

Concrete is currently the most common and most frequently used building material in different structural forms because of its strength and durability. Concrete is an acceptable building material and has some excellent properties, such as being rigid, low cost, and manufacturing ease [1]. Nevertheless, concrete has been commonly used for many years, it has undergone numerous improvements and enhancements during this time. Nowadays, concrete is not just a three-component mineral composite consisting of cement, aggregates, and water. Instead, it also includes mineral additives and chemical admixtures [2]. Many techniques have been used for enhancing the physical and mechanical properties of concrete by using auxiliary substances, including the incorporation of admixtures, supportive cement material (SCM), or other specific types of aggregates such as polycarboxylate polymer-based superplasticizers and varying water binder ratios [3]. The ecological effects of cement production can be minimized by the use of auxiliary substances, as a partial substituent of cement. The decreased use of cement as a binder in the concrete mixture has a positive influence on reducing

CO_2 emissions and reduces energy usage in the Portland cement production process [4]. Also, in the current scenario of large modern structures, the intrinsic weight needs to be reduced. Lightweight concrete is an innovative tool in this concern and has gained more publicity in recent years [5].

The use of lightweight concrete further boosts the building's structural capacity and reduces transportation and maintenance costs [6]. In general, lightweight aggregate concrete is obtained by mixing regular and lightweight aggregates which reduces the density of the aggregate. These aggregates include foamed slag, expanded shale, slag, vermiculite, expanded clay, pumice, perlite, or expanded polystyrene [7]. However, the manufacturing cost of lightweight aggregate concrete is more due to the involvement of a high-temperature sintering process. Also, the unequal distribution of aggregates renders the practical use of lightweight aggregate concrete. Aerated concrete is also known as foamed lightweight concrete, porous concrete or cellular concrete emerged as the modern construction material [8]. It has a relatively homogeneous and cellular structure as no coarse aggregates are used during its production. The cellular structure is because of the formation of air voids during the mixing process due to the introduction of a foaming agent. The materials used to manufacture the cellular concrete are the same as those used for traditional concrete, except the quartz aggregate and chemicals that produce the air voids [9]. Thus, the aerated concrete is a modification of normal concrete and the difference between them is in density and not quality. It exhibits properties like ordinary concrete with dense weight in most aspects, such as curing. The amount of air that can be included in aerated concrete precast or cast on-site varies from 20-50 per cent in volume in structural practices but may occupy 51 to 80 percent in concrete used for only thermal insulation, packaging, or filler purposes [10].

The presence of air voids in the cellular concrete imparts unique properties particularly, low density and lightweight. Cellular concrete provides the thermal and acoustic insulation properties depending on the density as the thermal conductivity increases linearly with the moisture content [11]. Cellular concrete has very low water absorption due to greater air content and high fire resistance. Its environmental friendliness, durability, and versatility increase the aesthetic value of buildings [12]. These distinctive properties of cellular concrete along with its easy production, high flow, and self-compacting nature result in its widespread use in numerous areas of civil and structural engineering such as the manufacture of pre-cast blocks and panel, shock-absorbing structures, road sub-base, and soil stabilization [13]. Its use in construction at regions under the threat of earthquakes and hurricanes is highly desirable. The physical and mechanical characteristics of cellular concrete are largely affected by the pore structure and porosity. The effect of foam content on the microstructure cement-based clay has been analyzed and an increase of permeability was obtained [14]. The curing duration and heat treatment have been found to significantly impact the pore structure and the performance of the silica fume based aerated concrete. The bulk density was also found to be dependent upon the content of the foaming agent [15]. The improvement in compressive strength has been observed with decreasing porosity in the presence of slag and an optimal water/binder ratio [16]. The pore distribution in the microstructure also affects the strength and durability of cellular concrete. The presence of large content of pores decreases the compressive strength but improves the thermal insulation and hygroscopic property as observed for foam concrete in the presence of silica fume and quick lime [17].

Foaming agents are used to lower the density of cellular concrete by generating and maintaining the air bubbles inside the cement paste [12]. The quality and quantity of the foaming agent pointedly affect the fresh and hardened properties of the cellular concrete. Various commonly used foaming agents include natural saponins and synthetic surfactants in addition to protein or resin-based polymeric compounds. The synthetic foaming agents have been found to provide more expansion with lesser density while the natural foaming agents provide a stable pore network structure [18]. The quality of foam is affected by various factors including water-binder ratio, the concentration of the foaming agent, mixing method, time of mixing, and curing type. The strength and rigidity of the resulting cellular concrete are strongly dependent upon the foam quality. The excessive use of the foaming agent has been reported to lead to a decrease in flow and compressive strength [19].

Prolonged mixing time reduces the air content and thereby increases the flow. The chemical admixtures with water reduction property have also been reported to lead to the destruction of the air network. The use of low content of foaming agents at a high water-binder ratio has been observed to decrease the consistency with increasing density of foamed concrete. On the contrary, the addition of superplasticizers has been found to increase flow [20]. The water-cement ratio has been found to significantly affect the microstructure and compressive strength of the cellular concrete [13].

The low mechanical strength of cellular concrete is a matter of concern and is usually dealt with by the incorporation of supportive cementitious materials (SCM) such as pozzolans [21]. It has been detected that the ultimate strength of foamed concrete is attained at a later age in the presence of high content of SCM [22]. These are siliceous substances with the composition of silico-aluminous or a combination of both. Most pozzolan materials are production residues of industrial or agricultural processes such as the burning of wastes left after agricultural production. These substances increase the density and impermeability, lower heat of hydration, and inhibit or counteract the expansions of cement matrix caused by the presence of free lime and free magnesia. The optimal incorporation of SCM has been found to result in intensification of the strength of cellular concrete. A significant impact of the alkali-activated fly ash has been observed on the heat resistance of cellular concrete [23]. The addition of an optimal amount of fly ash has been observed to improve the wetting strength, fluidity, and durability of cellular concrete [24]. These materials can replace between 15% and 40% of cement, without significantly reducing the strength and long-term durability of concrete. The most common materials used are fly ash, activated clays, foundry slag, and silica fume. The addition of carbon nanotubes has been found to enhance the uniform pore distribution and minimize the incurring strength loss [25]. Silica fume is another pozzolan that is generally used by replacing 5 to 10% Portland cement, and has the effect of reducing the average pore size of the capillaries in concrete; reduce its permeability; increase its resistance to chemical attack, and increase its compressive strength. Its role is two-fold, firstly it acts as an effective filler because of its fineness and, secondly chemically consumes the calcium hydroxide produced in the cement hydration to additionally produce calcium silicate hydrates (CSH gel). The effect of incorporation of foam in the presence of fly ash and silica fume has been observed on the physical and mechanical properties of cellular concrete at varying curing conditions [26]. The introduction of silica fume has been found to improve the microstructure and porosity of cellular concrete leading to enhanced strength [27].

Although literature reports many studies on fly ash based cellular concrete using various foaming agents, little data is available on the silica fume-based cellular concrete in the presence of polyurethane as a foaming agent. Hence, this study is aimed to obtain lightweight cellular concrete using polyurethane as a foaming agent at an optimum water-binder ratio in the presence of silica fume as supportive cementitious material. The effect of variation of percentage of polyurethane and silica fume for the cellular concrete has been studied. The study has been planned to explore the effects on the compressive strength, split tensile strength, and flexural strength at 7, 28, 56 & 90 days of curing ages of cellular concrete.

2. Materials and Methods

Ordinary Portland cement (43 Grade) was used. Silica fume was used as pozzolan and polyurethane was used as a foaming agent. These were obtained from the local supplier and were used as such without any modification. The Chemical composition of the binder materials have been given in Table 1. Natural sand conformed to IS: 383-1970 and a specific surface area of 256 m^2/kg was used as fine aggregates. The physical properties of SF have been given in Table 2. Polyurethane (PU) percentage was varied from 0-20% by volume while silica fume was mixed from 0-20% by weight of cement at a water-binder ratio of 0.35. The mix proportion has been given in Table 3. The components were mechanically mixed and moulded as per method given in literature [28]. The specimens were kept to set for 24 hours and then positioned in a water tank for curing at standard temperature. The strength analysis of the fully cured, surface dried concrete specimens was performed with a compression testing machine at 7, 28, 56 and 90 days as per IS 516-1959. Testing was carried out with three specimens for each mix and the average value was considered [29].

Table 1. Chemical Composition of Binder Materials

Oxides	SiO_2	Al_2O_3	Fe_2O_3	CaO	MgO	K_2O	Na_2O	TiO_2	SO_3	Cl	LOI
OPC (% by mass)	19.46	4.22	3.56	65.92	1.08	0.67	0.21	0.24	2.6	0.01	0.96
SF (% by mass)	98	0.03	0.02	0.3	0.2	0.3	0.3	-	-	-	0.8

Table 2. Physical Properties of SF

Property	Value
Size of Particle	<1.0 μm
Specific Gravity	2.23
Solid Bulk density	130–480 kg/m^3
Slurry Bulk density	1,450–1,600 kg/m^3
Densified Bulk density	560–720 kg/m^3
Specific Surface Area	13,200–26,500 m^2/kg

Table 3. Mix proportion of Specimens per m^3

Designation	Cement (kg)	SF (kg)	Water (litre)	PU (%)	Sand (kg)
CM	435	0	152.25	0	938
CC1	435	0	152.25	10	938
CC2	435	0	152.25	20	938
CCSF0	391.5	43.5	152.25	0	938
CCSF1	391.5	43.5	152.25	10	938
CCSF2	391.5	43.5	152.25	20	938
CCSF3	348	87	152.25	0	938
CCSF4	348	87	152.25	10	938
CCSF5	348	87	152.25	20	938

3. Results and Discussions

3.1. Density

Figure 1 illustrates the outcomes of the density determination of the concrete specimens. It can be observed that the density of the specimens decreases with the incorporation of PU, both in the absence of SF (CC1 and CC2) and presence of SF (CCSF0, CCSF1, CCSF2, CCSF3, CCSF4 and CCSF5). However, the decrease is small in the case of SF-based cellular concrete. Literature reports a decrease in the density of cellular concrete as related to conventional concrete due to the inclusion of air voids [15]. The comparative account of densities of control mix (CM) and SF-based cellular concrete reveals no significant impact on density in the absence of PU.

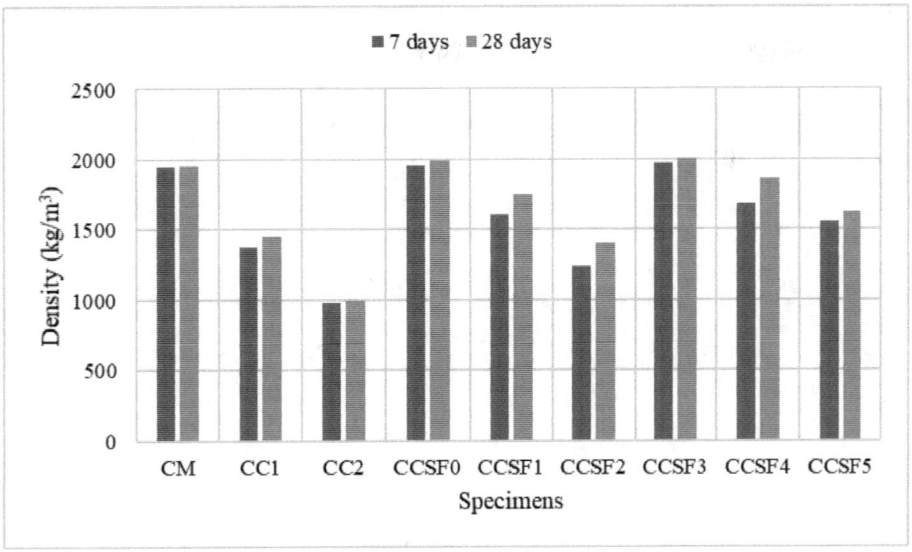

Figure 1. Density of Concrete Specimens

Figure 2 represents the effect of curing age with PU incorporation on the percentage decrease in densities of the cellular concrete specimens as related to the control mix. At 7 days of curing, the addition of PU decreased the density of CC1 by 29% and that of CC2 by 49% confirming that an increase in foaming agent decreases the density of cellular concrete. The results are consistent with earlier studies as literature reports the densities of cellular concrete varying from 770-1800 kg/m^3 [29]. Further, it has also been reported that the incorporation of pozzolans such as FA and SF improves the density of cellular concrete [5,17]. Also, the density is strongly affected by the fineness of the supportive cementitious materials. The fine silica particles form a coating over the air bubbles and prevent their flocculation resulting in uniform void structure. In the case of CCSF1 and CCSF2, the decrease in density was 18% and 36% respectively. Likewise, in the case of CCSF4 and CCSF5, a percentage decrease of 14% and 20% respectively was observed. However, as the curing age progressed, a slight increase in density was observed. The percentage decrease in density for CC1, CC2, CCSF1, CCSF2, CCSF4 and CCSF5 at 28 days was 26%, 49%, 11%, 28%, 5% and 17% respectively as compared to CM. The results were again consistent with earlier reported studies confirming the development of density with an increase in curing age [26].

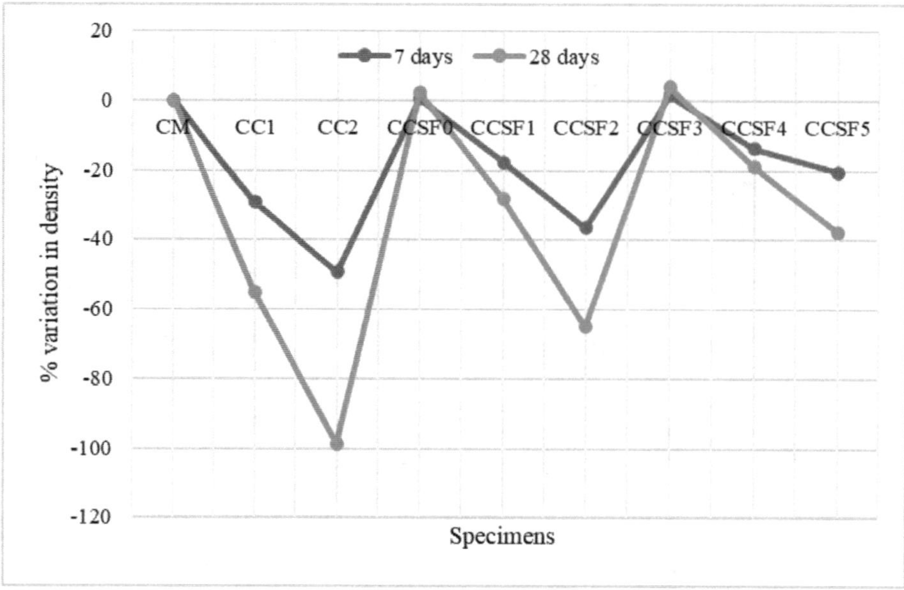

Figure 2. Variation in density of Concrete Specimens

3.2. Compression Strength

The outcomes of the compressive strength analysis of the concrete specimens have been represented in figure 3. The results reveal the significant impact of foaming agent on the compressive strength of cellular concrete. In comparison to CM, for CC1 and CC2, a decrease of 46% and 66% respectively was observed at 7 days of curing age, while a decrease of 44% and 64% was observed at 28 days. However, it is already reported that blends of foaming agents and pozzolans with finer material can improve the compressive strength of cellular concrete due to improved and uniform pore structure [5]. Better strength of limestone-based cellular concrete has been observed in the presence of a synthetic foaming agent [7]. Likewise, the incorporation of fly ash has resulted in better development of strength in cellular concrete [12]. Literature also reports similar observations in the presence of SF, where the introduction of SF progressed the compressive strength of cellular concrete significantly. The effect was pronounced at the optimal percentage of SF [26].

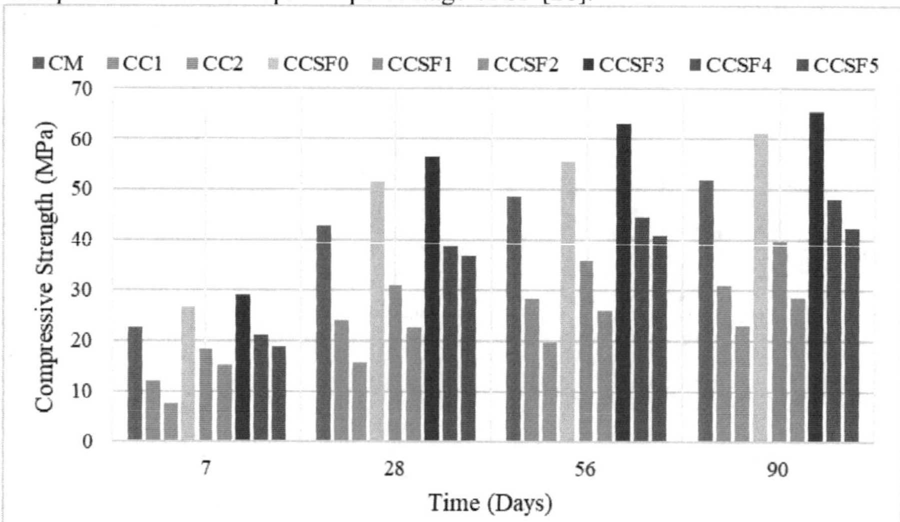

Figure 3. Compressive Strength of Concrete Specimens

A similar effect has been observed in the case of CCSF1 and CCSF2, where an increase of 19% and 32% respectively at 7 days and 28% and 47% increase at 28 days has been observed at the same content of PU as compared to CC1 and CC2. Further, the increase in SF content was observed to develop the compressive strength of the specimens. This effect is attributed to the amendment of pore structure leading to increased density and homogeneity of the matrix [18]. Hence, an increase of 6% and 67% for CCSF4 and 16% and 56% for CCSF5 respectively has been observed at 7 and 28 days. Since the introduction of SF results in the development of long-term strength of concrete, hence, the effect was more pronounced at later stages of curing. Thus, the decline in compressive strength due to decreased density of concrete in the presence of foaming agents can be managed by the addition of pozzolans such as SF. The pozzolanic action and filler effect of SF are primarily responsible for the strength development of concrete due to its participation in CSH gel formation [17].

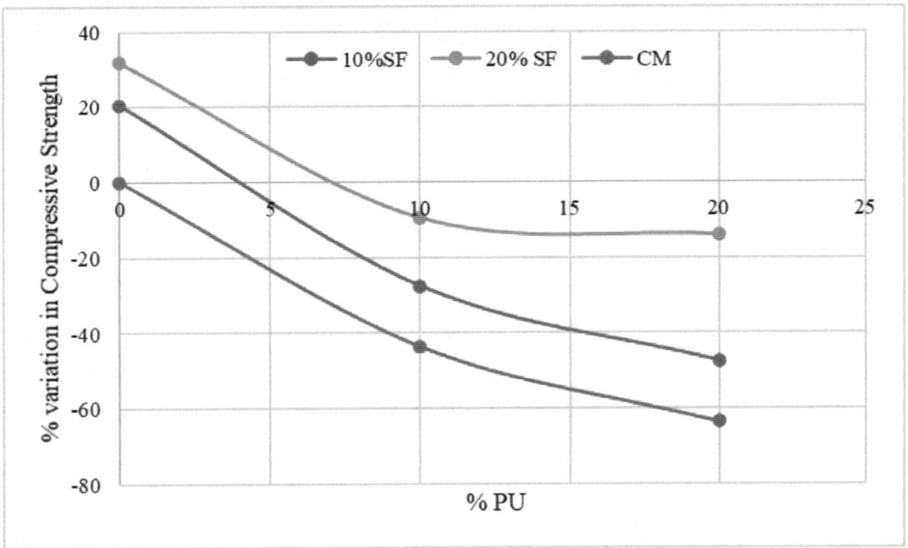

Figure 4. Effect of PU content on Compressive Strength of Concrete Specimens at 28 days

3.3. Split Tensile Strength

The outcomes of split tensile strength analysis were similar in pattern to that observed in the case of compressive strength analysis. Here also, a decrease in split tensile strength was observed with the incorporation of the foaming agent. Further, a decline in split tensile strength was observed with decreasing compressive strength of the specimen. Earlier studies on lightweight concrete also indicate similar behaviour and a correlation between split tensile and compressive strength [30]. Figure 5 illustrates the effect of incorporation of PU on the split tensile strength of concrete specimens in the presence and absence of SF at the varying volume of PU. In the case of CC1 and CC2, the strength decreased by 40% and 51% at 7 days and 43% and 55% at 28 days, respectively. Further, with an increase in the curing age of 7 to 90 days, a remarkable growth in split tensile strength was observed for all the specimens as illustrated in Figure 6.

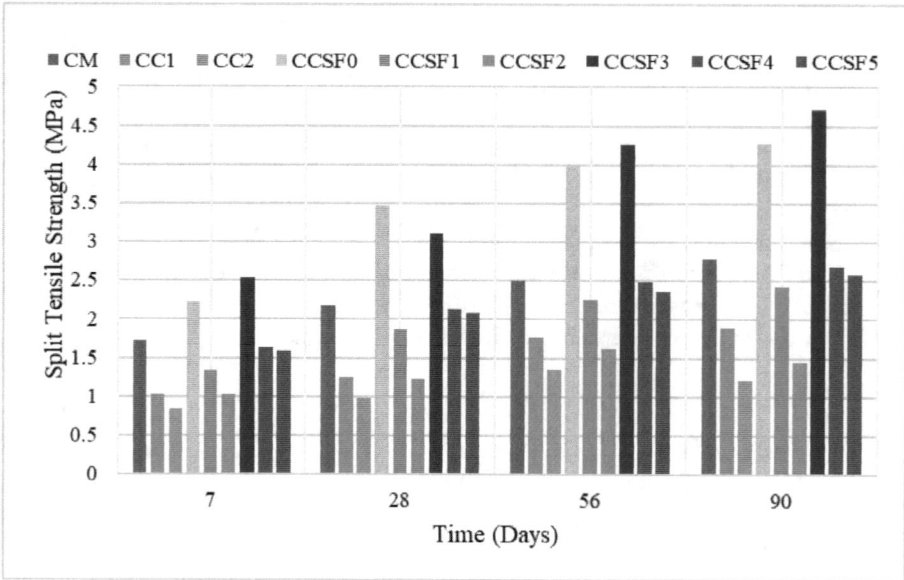

Figure 5. Split Tensile Strength of Concrete Specimens

The results indicate the increase in volume content of PU resulted in an increased porosity of the matrix leading to interfacial defects and load failure [8]. However, as the cement was partially substituted by SF, an increase in split tensile strength was observed. Thus, the % decrease for 20% SF was least as observed in figure. In the case of CCSF1 and CCSF2, an increase of 22% and 41% at 7 days and 14% and 44% at 28 days was observed. The effect was pronounced with an increasing SF percentage. The 7-day strength for CCSF4 and CCSF5 was higher by 5% and 8% respectively than the control mix specimen. The decrease at 28 days for CCSF4 and CCSF5 was observed as 2% and 4% respectively. The increase in splitting tensile strength is ascribed to the filler effect and pozzolanic activity of fine silica microparticles of SF that bring the better and refined distribution of pore structure [18]. Thus, a significant impact of the incorporation of SF on the splitting tensile strength of cellular concrete has been observed.

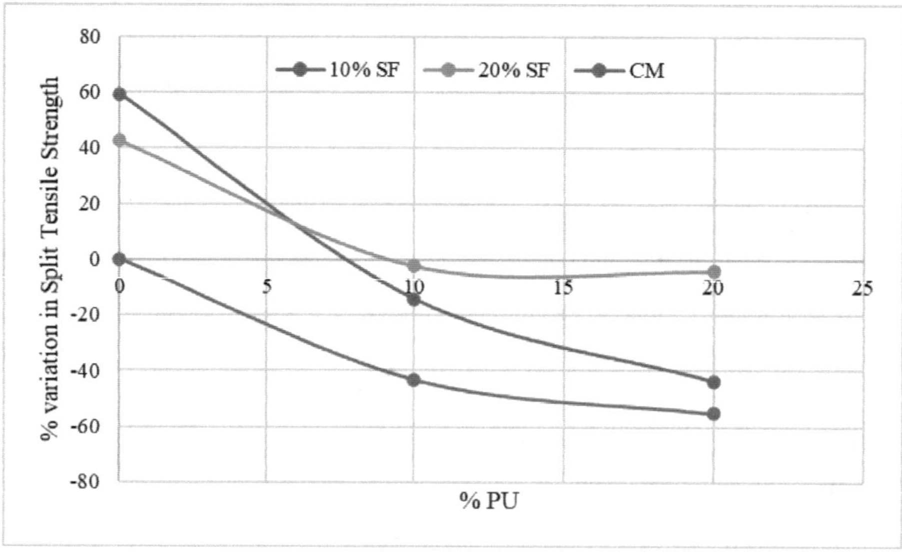

Figure 6. Effect of PU content on Split Tensile Strength of Concrete Specimens at 28 days

3.4. Flexural Strength

Figure 7 represents the behaviour of the flexural strength of the concrete specimens with the incorporation of foaming agent. Previous studies reveal a direct correspondence between compressive and flexural strength [9,10]. The flexural strength of cellular concrete specimens in the absence of SF was found to be lower as compared to the control mix. Hence, a decrease of 42% and 61% at 7 days along with a decrease of 37% and 53% was observed in flexural strength of CC1 and CC2 at 28 days respectively.

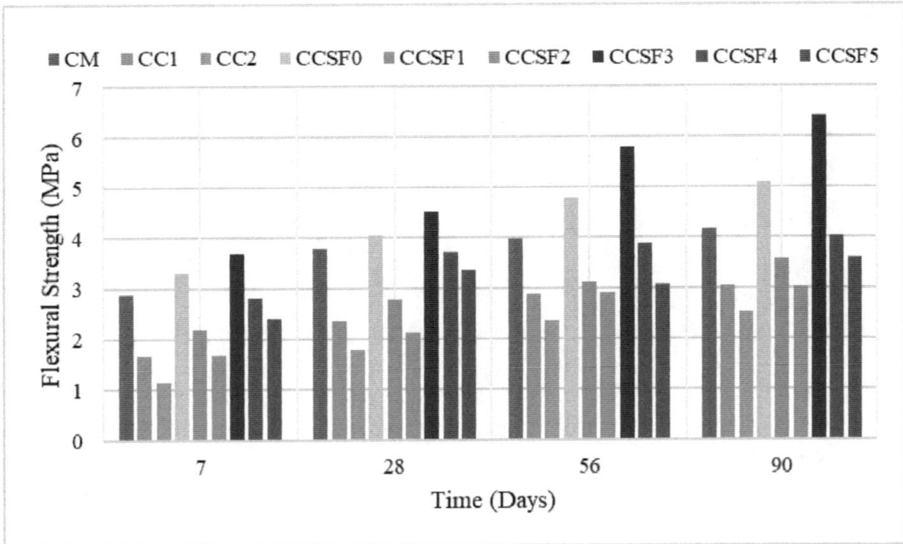

Figure 7. Flexural Strength of Concrete Specimens

The partial substitution of cement by SF increased the flexural strength of the specimens. In the case of CCSF1 and CCSF2, a decrease of 25% and 42% at 7 days, and 26% and 44% at 28 days was observed. The effect was more pronounced at increased content of SF. Thus, the flexural strength of CCSF4 and CCSF5 decreased by 2% and 17% at 7 days and 2% and 11% at 28 days, respectively. The test results show that at 20% content of SF and 10% of PU, the similar values of flexural strength were obtained for CCSF4 and CCSF5 as that of CM. Hence, the presence of pozzolanic SF provides a supportive impact on the matrix by improving the pore structure of cellular concrete [31,32].

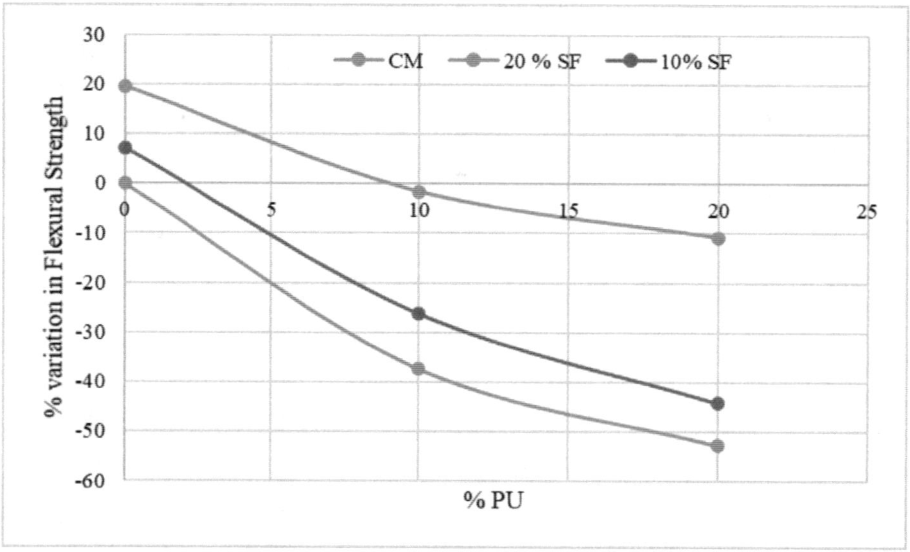

Figure 8. Effect of PU content on Flexural Strength of Concrete Specimens at 28 days

3.5. Correlation Analysis

Various researchers have reported correlation equations for the relationship between split tensile strength and compressive strength at various curing ages [7,18,30,32,33]. A correlation was obtained on a similar basis for concrete specimens with varying volume % of PU from 0-20% namely, CC (with the absence of SF), CCSF10 (with the presence of 10% SF), and CCSF20 (with the presence of 20% SF). The proposed model has been represented by equation 1.

$$f_{st} = af_c^{\,b} \qquad \text{(Equation 1)}$$

Where, f_{st} and f_c represent the split tensile strength and compressive strength respectively, while a and b are the model parameters. The values of the model parameters and correlation coefficient (R^2) have been listed in Table 3. The relationship between f_{st} and f_c has been represented in figure 8-10.

Table 4. Correlation Coefficients and Model parameters

7 days			28 days			56 days			90 days		
a	b	R^2	a	b	R^2	a	b	R^2	a	b	R^2
0.2035	0.6754	0.9754	0.104	0.8013	0.9763	0.1843	0.674	0.9983	0.0564	0.9957	0.9594
0.0231	1.3914	0.9993	0.0254	1.2485	0.9996	0.0325	1.1927	0.9944	0.0123	1.4253	0.9975
0.0508	1.155	0.9618	0.0646	0.96	0.9955	0.0118	1.4197	0.9876	0.0098	1.4713	0.9537

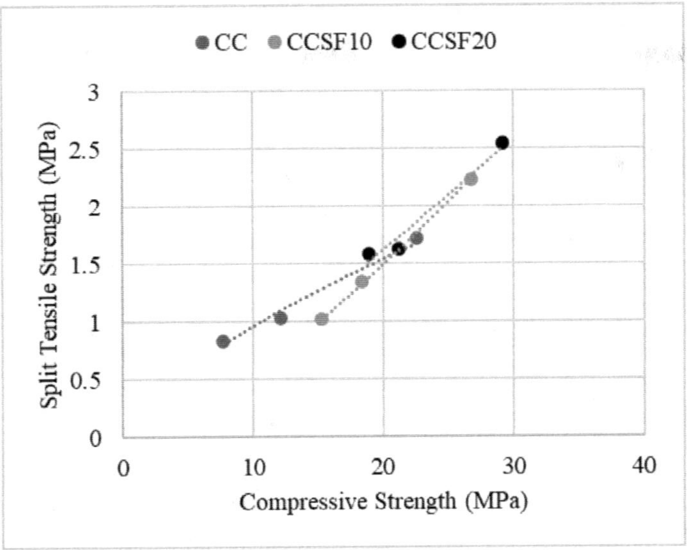

Figure 9. Correlation between Split Tensile Strength and Compressive Strength at 7 Days

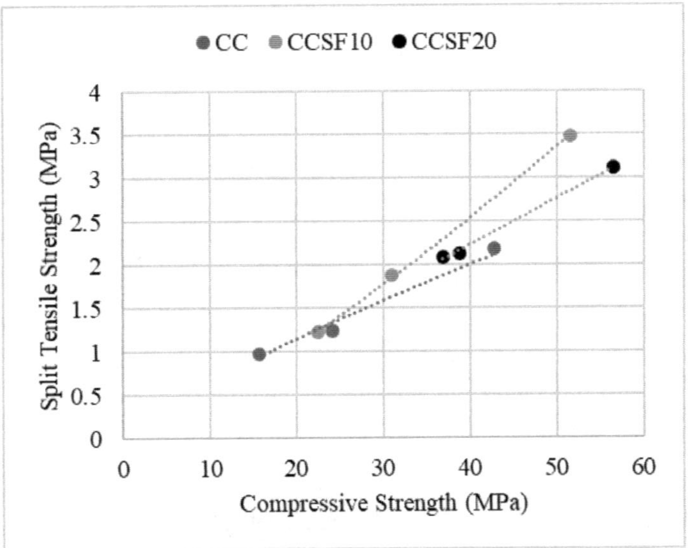

Figure 10. Correlation between Split Tensile Strength and Compressive Strength at 28 Days

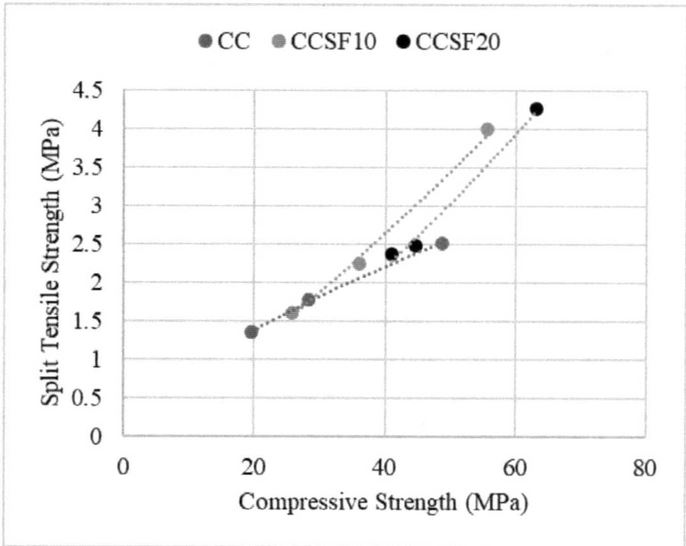

Figure 11. Correlation between Split Tensile Strength and Compressive Strength at 56 Days

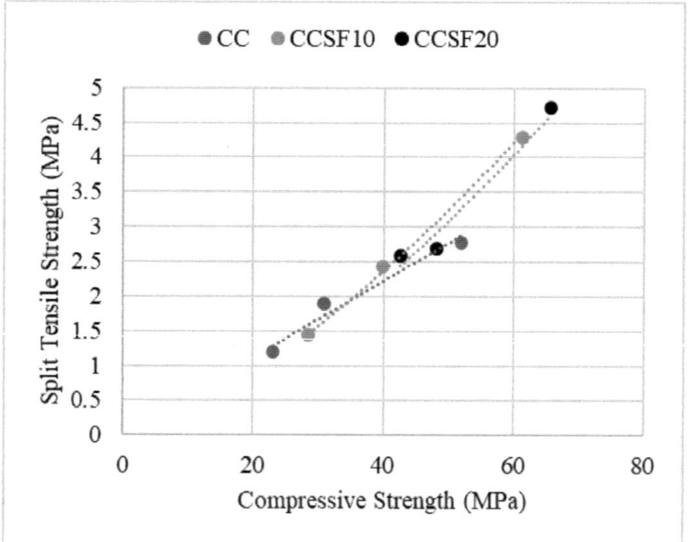

Figure 12. Correlation between Split Tensile Strength and Compressive Strength at 90 Days

4. Conclusion

Based on the above investigations, the resulting conclusions can be listed out:

1. The incorporation of foaming agents results in a significant decrease in density of the cellular concrete and the effect is pronounced with an increase in the content of the foaming agent. However, the inclusion of SF as a partial substituent of cement increased the density of the cellular concrete at the particular volume % of the foaming agent. The maximum increase was found for 10% inclusion of SF.
2. There was a comparative decrease in the compressive strength of the cellular concrete as compared to the control mix. The use of SF resulted in the enhancement of compressive strength due to its pozzolanic action as well as the ability to introduce homogeneity in the pore

structure. Similar effects were observed in the case of split tensile strength and flexural strength analysis.

3. There was a good correlation observed between split tensile strength and compressive strength for the studied days of treatment.
Hence, SF can be used to improve the performance of cellular concrete for further use in various construction purposes.

References

[1] Ashrafian A, Shokri F, Taheri Amiri M J, Yaseen Z M and Rezaie-Balf M 2020 Compressive strength of Foamed Cellular Lightweight Concrete simulation: New development of hybrid artificial intelligence model *Constr. Build. Mater.* **230** 117048

[2] Garg R, Bansal M and Aggarwal Y 2016 Strength, rapid chloride penetration and microstructure study of cement mortar incorporating micro and nano silica *Int. J. Electrochem. Sci.* **11**

[3] Bhosale A, Zade N P, Sarkar P and Davis R 2020 Mechanical and physical properties of cellular lightweight concrete block masonry *Constr. Build. Mater.* **248** 118621

[4] Garg R, Garg R, Bansal M and Aggarwal Y 2020 Experimental study on strength and microstructure of mortar in presence of micro and nano-silica *Mater. Today Proc.*

[5] Muthu Kumar E and Ramamurthy K 2017 Influence of production on the strength, density and water absorption of aerated geopolymer paste and mortar using Class F fly ash *Constr. Build. Mater.* **156** 1137–49

[6] Harith I K 2018 Study on polyurethane foamed concrete for use in structural applications *Case Stud. Constr. Mater.* **8** 79–86

[7] Colangelo F, Cioffi R, Liguori B and Iucolano F 2016 Recycled polyolefins waste as aggregates for lightweight concrete *Compos. Part B Eng.* **106** 234–41

[8] Tanyildizi H and Coskun A 2008 The effect of high temperature on compressive strength and splitting tensile strength of structural lightweight concrete containing fly ash *Constr. Build. Mater.* **22** 2269–75

[9] Mohammed B S, Foo W L and Abdullahi M 2014 Flexural strength of palm oil clinker concrete beams *Mater. Des.* **53** 325–31

[10] Wongkeo W, Thongsanitgarn P, Pimraksa K and Chaipanich A 2012 Compressive strength, flexural strength and thermal conductivity of autoclaved concrete block made using bottom ash as cement replacement materials *Mater. Des.* **35** 434–9

[11] Panesar D K 2013 Cellular concrete properties and the effect of synthetic and protein foaming agents *Constr. Build. Mater.* **44** 575–84

[12] Liu X, Ni C, Meng K, Zhang L, Liu D and Sun L 2020 Strengthening mechanism of lightweight cellular concrete filled with fly ash *Constr. Build. Mater.* **251** 118954

[13] Qiu Y, Li Y, Li M, Liu Y and Zhang L 2018 Experimental study on microstructure characters of foamed lightweight soil *IOP Conf. Ser. Earth Environ. Sci.* **108**

[14] Kikuchi Y, Nagatome T, Mizutani T A and Yoshino H 2011 The effect of air foam inclusion on the permeability and absorption properties of light weight soil *Soils Found.* **51** 151–65

[15] Chen Y L, Chang J E, Lai Y C and Chou M I M 2017 A comprehensive study on the production of autoclaved aerated concrete: Effects of silica-lime-cement composition and autoclaving conditions *Constr. Build. Mater.* **153** 622–9

[16] Wee T H, Daneti S B and Tamilselvan T 2011 Effect of w/c ratio on air-void system of foamed concrete and their influence on mechanical properties *Mag. Concr. Res.* **63** 583–95

[17] Cong M and Bing C 2015 Properties of a foamed concrete with soil as filler *Constr. Build. Mater.* **76** 61–9

[18] Bing C, Zhen W and Ning L 2012 Experimental Research on Properties of High-Strength Foamed Concrete *J. Mater. Civ. Eng.* **24** 113–8

[19] Nambiar E K K and Ramamurthy K 2006 Influence of filler type on the properties of foam

concrete *Cem. Concr. Compos.* **28** 475–80
[20] Jones M R and McCarthy A 2006 Heat of hydration in foamed concrete: Effect of mix constituents and plastic density *Cem. Concr. Res.* **36** 1032–41
[21] Garg R and Garg R 2020 Effect of zinc oxide nanoparticles on mechanical properties of silica fume-based cement composites *Mater. Today Proc.*
[22] Kearsley E P and Wainwright P 2001 the Effect of fly ash content on the compressive strength development of concrete *Cem. Concr. Res.* **31** 105–12
[23] Krivenko P and Kovalchuk G 2014 Achieving a heat resistance of cellular concrete based on alkali activated fly ash cements *Mater. Struct. Constr.* **48** 599–606
[24] Neramitkornburi A, Horpibulsuk S, Shen S L, Chinkulkijniwat A, Arulrajah A and Disfani M M 2015 Durability against wetting-drying cycles of sustainable lightweight cellular cemented construction material comprising clay and fly ash wastes *Constr. Build. Mater.* **77** 41–9
[25] Luo J, Hou D, Li Q, Wu C and Zhang C 2017 Comprehensive performances of carbon nanotube reinforced foam concrete with tetraethyl orthosilicate impregnation *Constr. Build. Mater.* **131** 512–6
[26] Gökçe H S, Hatungimana D and Ramyar K 2019 Effect of fly ash and silica fume on hardened properties of foam concrete *Constr. Build. Mater.* **194** 1–11
[27] Deng Z, Cheng H, Wang Z, Zhu G and Zhong H 2016 Compressive behavior of the cellular concrete utilizing millimeter-size spherical saturated SAP under high strain-rate loading *Constr. Build. Mater.* **119** 96–106
[28] Liu X, Wu T, Yang X and Wei H 2019 Properties of self-compacting lightweight concrete reinforced with steel and polypropylene fibers *Constr. Build. Mater.* **226** 388–98
[29] Vinith Kumar N, Arunkumar C and Srinivasa Senthil S 2018 Experimental Study on Mechanical and Thermal Behavior of Foamed Concrete *Mater. Today Proc.* **5** 8753–60
[30] Zhang S, Cao K, Wang C, Wang X, Deng G and Wei P 2020 Influence of the porosity and pore size on the compressive and splitting strengths of cellular concrete with millimeter-size pores *Constr. Build. Mater.* **235** 117508
[31] Falliano D, De Domenico D, Ricciardi G and Gugliandolo E 2018 Experimental investigation on the compressive strength of foamed concrete: Effect of curing conditions, cement type, foaming agent and dry density *Constr. Build. Mater.* **165** 735–49
[32] Shafigh P, Jumaat M Z, Mahmud H Bin and Hamid N A A 2012 Lightweight concrete made from crushed oil palm shell: Tensile strength and effect of initial curing on compressive strength *Constr. Build. Mater.* **27** 252–8
[33] Wang X hua, Zhang S rong, Wang C, Cao K lei, Wei P yong and Wang J xin 2019 Effect of steel fibers on the compressive and splitting-tensile behaviors of cellular concrete with millimeter-size pores *Constr. Build. Mater.* **221** 60–73

IOP Conf. Series: Materials Science and Engineering **961** (2020) 012009 doi:10.1088/1757-899X/961/1/012009

Mechanical Properties and Durability Study of Jute Fiber Reinforced Concrete

S Tiwari[1], A K Sahu[2] and R P Pathak[3]

[1] Assistant Professor, Department of Civil Engineering, GLBITM, Greater Noida, India
[2] Assistant Professor, Department of Civil Engineering, RRIMT, Lucknow, India
[3] Scientist C, Central Soil and Materials Research Station, New Delhi, India

E-mail: sachinkbtechno@gmail.com

Abstract. Natural material/fiber should be used in the construction industry as it finds low cost and improve the properties of the material. Jute fiber is used in research study and carried out an experimental investigation on the mechanical properties of the jute fiber reinforced concrete (JFRC). Natural available jute fiber was chopped to the desired length and it was mixed in concrete to produce JFRC. The chopped jute fiber added in three different percentages i.e. 0.5%, 1.0%, & 1.5% in three various concrete mixes (M25, M30 and M40). Additionally, JFRC concrete specimens cured in the acid medium and examine the compression strength, split tensile strength, and strength reduced under acid curing. Workability results indicated that the slump value (workability) reduced as an increased amount of jute fiber in the concrete specimens. Also, the compressive strength reduced in the acid curing as compared to normal curing. Additionally, Jute fiber increased the compressive and tensile strength of every concrete mix. This research study revealed that natural fiber (jute fiber) can be used as additives to enhance the durability and strength of concrete.

Keywords: *Concrete mix, Jute fiber, Compressive strength, Tensile strength, Acid curing, Workability*

1. Introduction

Natural fibers are cost-effective with low density, biodegradable, renewable, on-abrasive and readily available worldwide. The natural fiber is mainly composed of cellulose, hemicelluloses and lignin with minor amounts of protein and inorganic. Jute Fiber Reinforced Concrete (JFRC) can be considered as a composite material made of concrete and short discrete natural jute fiber. Therefore, much research has been undergoing to utilize the naturally available jute fiber to use in a concrete matrix, for creating a composite building material. The development of such composite material will increase the strength of the structure and also increases the life span of the structure. Many studies carried out on the concrete mix with addition of jute fiber. Aziz et al. [1] studied the critical estimation and properties of natural fiber reinforced concrete and it was seen that the suitable technology for quality product development was of main importance. Ramakrishna et al. [6] worked on the four different fibers such as coir, sisal, jute and Hibiscus cannabinus using in mortar mix. Results found that coir fiber was significantly improved the initial strength. Sen and reddy [8] investigated the materials chosen for structural up-gradation must, in addition to functional efficiency and increasing or improving the various properties of the structures, should fulfill some criterion, for the cause of sustainability and a better quality.

Zhou et. al. [10] researched fracture and impact properties of short discrete jute fiber reinforced cementitious composites (JFRCC) with various form for developing low-cost natural fiber-reinforced concretes and mortars for construction. Three different process methodologies were adopted to mix the jute fiber homogeneously in the mortar matrix [2]. Out of these three processes, the best dispersion of the fibers was achieved when water-saturated chopped jute fibers were used in mortar sample preparation. Additionally, Oh et al. [5] worked on the jute fiber reinforced concrete with an addition three different percentage of jute fiber. Alkali activator was also used and replaced the blast furnace slag. With the increased amount of jute fiber and alkali activator, void ratio and compressive strength was also increased. Yan et al. [9], studied the flexural strength using two types of concrete such as plain concrete and coir fiber reinforced concrete (CFRC). Flax fabric reinforced epoxy polymer (FFRP) was applied on the beam specimens. FFRP layer and coir exhibited the flexural strength of specimens. Razmi and mirsayar [7] used 20mm length fiber with three different percentages of 0.1%, 0.3% and 0.5% and evaluate the toughness and mechanical properties of concrete. The results found that concrete mix with jute fibers improved the compressive, tensile and flexural strength as well as higher resistance value. Additionally Islam and ahmed [4] used three different dosage of jute fiber i.e. 0.25%, 0.5% and 1% with two length of 10 mm and 20 mm. Maximum improvements was observed in jute fiber concrete at the addition of 0.25% jute fiber.

Various research studies were performed on the concrete and mortar mix using jute fiber with different percentages. Therefore, a study needs to be done to examine the effect of chemical/acid on the concrete mix with jute fibers. This research study aims to evaluate the effect of jute fiber (chopped) on the compressive strength and tensile strength of the concrete after 7 days and 28 days. Jute fiber was used with three different percentages of 0.5%, 1% and 1.5% in three concrete mixes (M25, M30 and M40) to prepare the Jute fiber reinforced concrete (JFRC). Also, concrete specimens with jute fiber cured in the acid medium to examine the effect of acid on the mechanical properties of concrete specimens after 28 days.

2. Material used and methods

2.1 Materials

All the materials used in the present study was tested under the IS guidelines. Ordinary Portland cement (OPC-43) was used in the research study. The specific gravity and consistency of the cement was 3.14 and 34% respectively. The compressive strength of the cement was 42.6 MPa after 28 days. The specific gravity, water absorption and fineness modulus of the coarse aggregate (CA) and fine aggregate (FA) was 2.72, 0.7%, 6.9 and 2.48, 1%, 2.44 respectively. Grading zone of the coarse aggregate and fine aggregate lied in II zone. The admixture was also used in this study. CICO C -300 was used as water reducing admixture in the concrete mix as confirmed by IS 9103:1999.

2.2 Jute fiber

Jute fiber is a three-dimensional composite mainly composed of cellulose, hemicelluloses and lignin with minor amounts of protein and inorganic [7]. Jute fiber has high tensile strength as compared to other fiber like sisal, bamboo, coir, & hemp. Apart from the tensile strength jute fiber can easily withstand heat. The different mechanical properties of the jute fiber has-been briefly described here in Table 1. Raw jute fiber (Figure 1(a)) collected from the manufacturer and chopped (Figure 1(b)) in the concrete laboratory.

Table 1: Properties of jute fiber

Properties	Values
Specific Gravity [kg/m^3]	1460
Water Absorption [%]	13
Tensile Strength [MPa]	400-800
Stiffness [kN/mm^2]	10-30

Figure 1 (a) Raw jute fiber (b) Chopped jute fiber

2.3 Mix Design of three different grade of concrete

Concrete mix design of M25, M30 & M40 is prepared as per IS: 10262-2009 [3]. These grades are selected because beams and column are mainly cast of selected concrete mix proportions on construction site. Several trial mixes has been done to obtain the optimum design mix of concrete for different grades then add the jute fibers. The jute fiber has been added in three different percentages into the cement matrix i.e. 0.5%, 1.0%, & 1.5%. The procedure followed for preparing mix design is given in Table 2 for different grades of concrete mix.

Table 2 Different mix proportions of concrete

Materials	Mix proportions (kg/m^3)		
	M25	M30	M40

Cement	315.44 kg	366.79 kg	415.05 kg
Water	157.72	157.72	157.72
Fine aggregate (F.A.)	735 kg	719 kg	704 kg
Coarse aggregate (C.A.)	1200.9 kg	1174.3 kg	1149.3 kg
Superplasticizer	3.15	3.67	4.15
Water- cement (W/C) ratio	0.50	0.43	0.38

2.4 Preparation and testing of specimens

All the experimental work in the present research has been carried out as per the IS codes. Local available jute fiber was taken in the study and raw jute fiber was chopped up to the length of 25-50mm in length for the preparation of the JFRC. The chopped jute fiber was immersed in water for 24 hour before mixing it in the concrete matrix. The jute fiber was mixed in the concrete, after preparing the cement and jute slurry. 50% of cement and water were taken with the required amount of jute fiber for preparing the slurry. The remaining 50% dry cement mixed with coarse and fine aggregate in the mixture. Once the dry mix properly mixed, the slurry was added into the mixture, and remaining 50% of water with superpasticizer was added. This procedure was selected for the proper mixing of jute fiber with materials as well as jute fiber did not soak the additional water of concrete. In this way, the Jute fiber reinforced concrete was prepared for the three different grades of concrete.

Freshly prepared concrete was then poured in the moulds. The cube and cylinder specimens were de-moulded after 24 hours and were kept for a curing period of 7 and 28 days. The concrete cube and cylinder specimens cured in the acid medium (Sulphuric acid 0.5N) for 28 days as various concrete structures is effected by chemical attack. Also, see the impact of sulphuric acid on jute fiber concrete. Cube and cylinder specimens were prepared to evaluate the compressive and tensile strength of concrete using compression testing machine.

3. Result and discussion

Destructive test were conducted on hard concrete after 7 days and 28 days of curing. Whereas, on freshly prepared concrete slump test was done to check the workability of the jute fiber reinforced concrete. To know the effect of acid curing (sulphuric acid) on concrete, a destructive test was conducted after 28 days of acid curing. The various destructive test has been done on the hardened concrete to check the compressive and split tensile strength of the concrete.

3.1 Workability test results

A slump cone test was done on the freshly prepared concrete to know the workability (slump value) of the concrete mix. The slump value was 110mm, 105mm and 110mm for the conventional concrete grade of M25, M30 and M40 respectively. Less value of slump value (80mm) was observed at 1.5% percentages of jute fiber in M25 and M40 grade of concrete. The slump value was decreased as the percentages of jute fiber increased in the concrete mix due to the two factors. The first factor is when the chopped jute fiber mixed into the concrete matrix the jute got agglomerated which result in non-uniform distribution of jute. Secondly, jute being hydrophilic, it absorbs water from the mix which required to make concrete, hence reduces the workability. Figure 2 shows the bar chat of slump value reduction with the addition of jute fiber into the cement matrix.

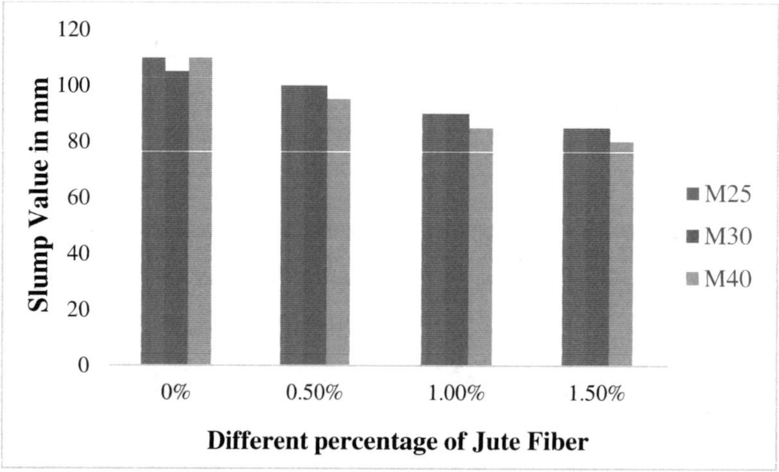

Figure 2 Slump values of control mix and JFRC

3.2 Compressive strength test results

Testing of hardened concrete is important for determination of mechanical properties of concrete as it can give a reasonable estimation of the fundamental physical behavior of concrete such as compressive strength, hardness and homogeneity of concrete for quality

IOP Conf. Series: Materials Science and Engineering **961** (2020) 012009 doi:10.1088/1757-899X/961/1/012009

control aspect.

The compressive strength results of the cube specimens with the addition of different percentage of jute fiber for concrete grade M25, M30 and M40 after 7 days & 28 days under normal and acid curing is presented in Figure 3, Figure 4 and Figure 5 respectively. The compressive strength result of control mix was 20.3 MPa, 26.8 MPa and 21.35 MPa at 7 days, 28 days and in acid curing respectively. Highest compressive strength was observed as 23.2 MPa, 32.1 MPa and 21.20 MPa after 7 days, 28 days and in acid curing respectively at 1.5% addition of jute fiber however lowest compressive strength was observed as 20.9 MPa, 27.7 MPa and 21.20 MPa at 7 days, 28 days and in acid curing respectively at 0.5% addition of jute fiber. The compressive strength of the concrete increased with the addition of jute fiber into the concrete matrix. As fiber bridges the micropores and fills the voids in the concrete which helps to enhance the strength of concrete. The present study revealed that maximum strength was achieved as 19.7 percentage increased with the addition of 1.5% jute fiber into the concrete matrix for M25 grade of concrete after 28days of curing due to proper uniformity distribution of fiber inside in the concrete matrix. Whereas the percentage in reduction of strength after 28 days acid curing, was maximum at 1.5% jute fiber, which is found to be 37.8%.

The compressive strength values were 26 MPa, 34 MPa and 28.4 MPa of the control mix (M30) at 7 days, 28 days and in acid curing. The maximum compressive strength was observed as 29.1 MPa and 37.3 MPa after 7 days and 28 days which is found to be increased by 11.9% and 9.7% respectively. The strength of concrete keeps on increased with addition of fiber as shown in Figure 4.3. Whereas the reduction of strength after 28 days in acid curing, was maximum at 1.5% jute fiber, which is found to be 31.06%. The results indicated that addition of jute fiber slightly increased the compressive strength in acid curing up to 1% of jute fiber however; it was reduced at 1.5% of jute fiber.

A similar trend of results was observed in the M40 grade of concrete as observed in M25 and M30 grade of concrete. Figure 4.4 shows the results for the M40 grade of concrete. 27.2 MPa, 44.1 MPa and 36.70 MPa strength was found in the control mix at 7 days, 28 days and in acid curing after 28 days. The maximum increase in strength after addition of jute fiber was achieved at 1.5% which was found as 35.1 MPa and 47.7 MPa at 7 days and 28 days respectively. The compressive strength was observed minimum at 0.5% addition of jute fiber.

Acid curing showed negative impact on the strength of concrete and the compressive strength decreased in the acid curing as compared to normal curing.

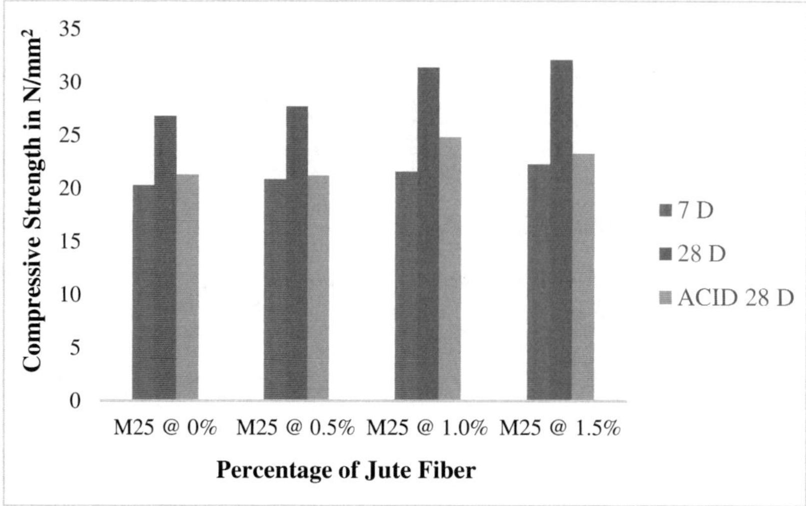

Figure 3 Compressive strength results of control mix and JFRC for concrete (M25)

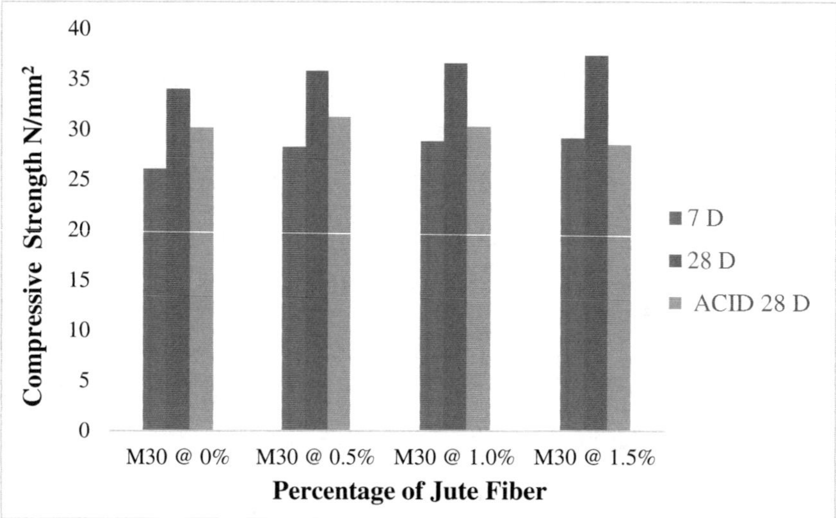

Figure 4 Compressive strength results of control mix and JFRC for concrete (M30)

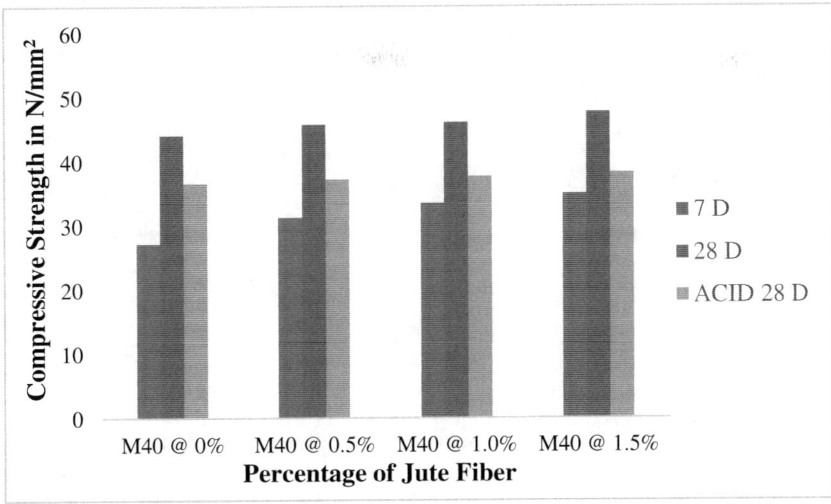

Figure 5 Compressive strength results of control mix and JFRC for concrete (M40)

3.3 Split tensile strength result

Split tensile test has been conducted to observe the brittle nature of the concrete specimens that consisted of natural aggregate and jute fiber in differing percentage. The split tensile test is carried out on cylinder specimens (150mm diameter, 300 mm depth) according to IS: 5816-1999.

The concrete mix with 0% jute fiber (CM) showed tensile strength as 2 MPa, 2.6 MPa and 2.1 MPa at 7 days, 28 days and in acid curing as presented in Table 3. Concrete mix with fiber dosage at 1.5%, achieved maximum strength as 3.4 MPa which was found to be increased by 30.8% as compared to control sample. However, the mixes with fiber dosage of 0.5% and 1% showed minimum tensile strength as 3 MPa and 3.3 MPa respectively after 28 days. The split tensile strength increased with the addition of jute fiber, which shows that the brittle nature of concrete can be overcome by addition of jute fiber for M25 Grade of concrete. This improvement in the tensile strength of JFRC was observed due to the jute fiber. Besides, the cylinder specimens cured in the acid curing to examine the tensile strength. The tensile strength gradually decreased in acid curing as compared to normal curing because acid medium accelerates the rate of decomposition for the jute fiber. Additionally, it is a natural fiber which has the natural tendency to undergo decomposition with time.

Similar observations of tensile strength were obtained in M30 and M40 grade of concrete. Incorporation of jute fiber with three different percentages in both concrete mixes (M30,

IOP Conf. Series: Materials Science and Engineering **961** (2020) 012009 doi:10.1088/1757-899X/961/1/012009

M40) increased the tensile strength after 7 days and 28 days. Also, jute fiber increased the tensile strength of cylinder specimens in acid curing but not significantly improved as compared to normal curing. The tensile strength was slightly low in acid curing as compared to normal curing. The maximum increment in strength was 19.35% (3.7 MPa) as compared to the control mix at 1.5% content of jute fiber for M30 concrete. The strength in acid curing showed reduction in strength was minimum at 0.5% jute addition, which was found to be 26.92% (2.6 MPa) as presented in Table 4. For the M40 grade of concrete, highest tensile strength was observed as 4.6 MPa (21.1%) at 1.5% jute fiber however lowest strength was observed as 4.1 MPa (7.9%) at 0.5% jute fiber at 28 days presented in Table 5. 2.97 MPa strength (1.5% jute fiber) was observed maximum in acid curing. Inclusion of jute fiber content (0.5%, 1%, and 1.5%) in the concrete mixes did not effectively increased the tensile strength of cylinder specimens cured in acid medium after 28 days as compared to normal curing.

Table 3 Split tensile strength result of JFRC concrete (M25)

Grade with Jute fiber percentage	**Strength after 7 D**	**% Increase**	**Strength after 28 D**	**% Increase**	**After Acid Curing**	**% Reduction after Acid Curing**
M25 @ 0%	2.0	-	2.6	-	2.1	23.8
M25 @ 0.5%	2.2	10	3.0	15.4	2.3	30.43
M25 @ 1.0%	2.4	20	3.3	26.9	2.6	26.92
M25 @ 1.5%	2.5	25	3.4	30.8	2.97	30.76

Table 4 Split tensile strength result of JFRC concrete (M30)

Grade with jute fiber percentage	Strength after 7 D	% Increase	Strength after 28 D	% Increase	After Acid Curing	% Reduction after Acid Curing
M30 @ 0%	2.4	-	3.1	-	2.5	24
M30 @ 0.5%	2.7	12.5	3.3	6.4	2.6	26.92
M30 @ 1.0%	2.9	20.8	3.6	16.12	2.78	29.5
M30 @ 1.5%	3.1	29.1	3.7	19.35	2.8	32.14

Table 5 Split tensile strength result of JFRC concrete (M40)

Grade with jute fiber percentage	Strength after 7 D	% Increase	Strength after 28 D	% Increase	After Acid Curing	% Reduction after Acid Curing
M40 @ 0%	2.5	-	3.8	-	3.1	22.5
M40 @ 0.5%	3.0	20	4.1	7.9	3.3	24.2
M40 @ 1.0%	3.3	32	4.3	13.15	3.3	30.3
M40 @ 1.5%	3.4	36	4.6	21.1	3.5	31.42

4. Conclusion

Jute fiber has been added in three different percentages i.e. 0.5%, 1.0%, & 1.5% in M25, M30, and M40 grade of concrete for making concrete better in its mechanical property. Furthermore, the durability of concrete is studied to know the effect of acid curing on JFRC. While studying, every aspect has been deeply observed and following conclusions are deduced from the present investigation:

- The workability of the JFRC reduced with the addition of jute fiber in the concrete matrix. The maximum workability was achieved at 0% of jute fiber addition in all grades of concrete. Whereas minimum workability was observed with 1.5% addition of jute fiber.

This reduction in workability is due to hydrophilic nature of jute fiber. The jute fiber absorbs the water from the concrete mix which is required for its strength, heat of hydration and workability. Therefore, the JFRC mix had less water content, which reduced the workability.

- The compressive strength M25, M30 & M40 of concrete is increased with the addition of jute fiber. Maximum compressive strength was achieved at 1.5% addition of jute fiber in M25, M30 & M40 grade of concrete which amount to be 19.7%, 9.7% and 8.1% respectively. However, the compressive strength of specimen cured in acid medium showed maximum reduction in strength was also observed at 1.5% addition of jute fiber. This is due to the natural decay of jute fiber and one major drawback of jute fiber as the JFRC specimens placed in acid medium so it starts to accelerate the decomposition process of jute fiber.
- The split tensile strength results were similar to the compressive strength of concrete specimens and maximum tensile strength was obtained in all concrete mixes after 28 days of curing at 1.5% jute fiber content. Improvement in tensile strength was observed as 30.8%, 19.35% and 21.1% in M25, M30 and M40 grade of concrete respectively. Whereas, the maximum reduction in strength is observed for JFRC with 1.5% addition of jute fiber after 28 days in acid curing.
- Jute fiber binds the materials inside into the concrete matrix and enhances the compressive and tensile strength of concrete. However, JFRC did not show any significant results in acid curing and cannot use this concrete in coastal/marine areas.

References

[1] Aziz M A, Paramasivam P, and Lee S L 1981 *Int. J. Cem. Compos. Light Weight Concrete* **3** 123-132.

[2] Chakraborty S, Kundu P S, Roy A, Basak K R, Adkhikari B, and Majumder S B 2013 *Constr. Build. Mater.* **38** 776-784.

[3] IS 10262 2009 *Bureau of Indian Standards* New Delhi India.

[4] Islam S M, Ahmed J U S, Zhou X, Ghaffar H S, Dong W, Oladiran O, and Mizi F 2018 *Constr. Build. Mater.* **189** 768–776.

[5] Oh R, Cha S, Park S, Lee H, Park S, and Park C 2014 *Paddy Wat. Environ.* **12** 149–156

IOP Conf. Series: Materials Science and Engineering **961** (2020) 012009 doi:10.1088/1757-899X/961/1/012009

[6] Ramakrishan G and Sundararajan T 2016 *Cem. Concr. Compos.* **27** 547-53.

[7] Razmi A and Mirsayar M M 2017 *Constr. Build. Mater.* **148** 512-520.

[8] Sen T and Reddy J N H 2011 *Int. J. Innov. Manag. and Technol.* **2** 186-191

[9] Yan L, Shen S, and Nawawi C 2015 *Part B: Engg.* **80** 343-354.

[10] Zhou X, Ghaffar H S, Dong W, Oladiran O, and Mizi F 2013 *Mater. Des.* **49** 35-47.

IOP Conf. Series: Materials Science and Engineering **961** (2020) 012010 doi:10.1088/1757-899X/961/1/012010

Performance analysis of Papercrete in presence of Rice husk ash and Fly ash

A Singh[1], S Singla[1*], R Garg[2], and R Garg[3]

[1]Department of Civil Engineering, RIMT University, Mandi Gobindgarh, India
[2]Department of Civil Engineering, Galgotias College of Engineering and Technology, Greater Noida, India
[3]Department of Chemistry, Rayat Bahra University, Mohali, India
*Corresponding author: drsandeepsinglaz@gmail.com

Abstract. Papercrete is a modern composite material, that uses waste paper as a partial replacement to Portland cement, and is a renewable medium to construction owing to the reduction of the volume of waste paper as well as cement. Papercrete has strong sound absorption, thermal insulation and a low-cost alternative; light weight and fire-resistant building material. However, its low mechanical strength needs to be compensated by incorporation of supplementary cementitious materials. The work discusses the potential for the partial replacement of cement (by weight) with waste paper pulp in presence of Rice husk ash (RHA) and fly ash (FA) in ternary blends. For tests, the levels of replacement for RHA and FA were 10%, 15%, 20% and 25% and for waste paper, the rates of replacement were maintained at 10%. A reference mix was also prepared for the comparative purposes without substitution by RHA, FA and paper pulp. The flow, compressive strength, split tensile strength, water absorption and dry density studies were performed for all the specimens. The findings revealed decreased the flow of specimens compared with the reference mix along with reduction in strength that was well compensated by RHA and FA particles. The optimal content for RHA blends was 20% and that of FA was 15% with maximum strength among the mixes. The study signifies that the waste paper can be used to prepare papercrete in presence of optimal content of supplementary cementitious materials to have desired mechanical properties.

1. Introduction

The emerging market for building materials needs to satisfy all the challenging demands for short-term and long-term structural efficiency [1]. The building materials of tomorrow will have to meet more stringent performance standards than those in practice today, as they will be more elaborative and more complex [2]. A huge amount of money is required for large housing construction activity in developing countries. Building materials account for about 70% of the total cost to build housing [3]. The growing popularity in the construction industry of renewable, inexpensive and lightweight building materials has led us to explore ways to accomplish this through the conservation of the environment and the preservation of standards-supported material requirements [4]. Bricks are the most frequently used building materials in building construction worldwide as one of the oldest building materials [5]. The bricks are moulded into rectangular blocks in uniform size, and the blocks are dried and burned. The bricks can be arranged correctly as they are uniform in size. But it does have two main disadvantages, self-weight and fragility. The brick-making method has not changed in India for decades, except for a few minor refining steps. In our country, there was little effort to improve the process of making bricks to improve brick quality. The structures were also not drastically modified in view of their compressive strength, structural stability and relatively low costs [6].

Initially developed 80 years ago, Papercrete was only recently re-discovered as a small-scale concept. Papercrete is a cement fiber paste made of Portland concrete and waste paper combined with water to produce a pulp of paper cement, which can then be poured into a mold to dried and then used for long term building [7]. The processing of Papercrete does not contain hazardous by-products or an unnecessary use of energy. Papercrete can be either fibrous, padobe or fidobe type. A mixture of paper, portland cement and water is a fibrous papecrete. Padobe has no cement and is a blend of paper and water in presence of clay as a binding medium [8]. The standard brick can crack during drying when the clay content is too high, but with paper fibres in the earth mix the drying block strengthens. This offers stability to avoid cracking. Fidobe is like padobe, but other fibrous material may be found there. Paper is a natural polymer consisting of the most plentiful organic compound in the planet i.e. wood cellulose. Cellulose is water insoluble, although it contains many hydroxyl groups. The reason is because the chains are rigid and hydrogen bonds are present in adjacent chains between two OH groups [8]. The chain gets coiled consistently in places, forming strong, stable crystalline regions that increase stability and strength in coiled chains. The strength of papercrete is founded on this hydrogen-bonding. Applying force on the surface, breaks the hydrogen bonding between the water and cellulose molecule. Coating of Portland cement on cellulose fibres produces a cement matrix to give the blend extra strength [9].

The use of industrial waste of various forms for production of building materials is a new era for building technology in the construction industry. In the last few years waste rubbers, glass powders, fibres, waste sawdust and calcareous waste in powder have been taken into greater account for manufacturing buildings [6]. This is fully comprehensible because the economic progress of construction is becoming more and more recognized as more dependent on intelligent use of materials and constant improved materials. In the last decade, a great demand has been placed on the construction materials industry because of the increasing population which causes a chronic shortage of building materials [10]. The situation has posed a challenge to civil engineers to turn industrial wastes into usable construction and building materials. Increased environmental concern also resulted from an accumulation of unprocessed waste in developing countries [11]. In addition to the problems of pollution, the economic design of buildings can be dealt by reusing waste in building material as a viable option. The existing knowledge pool of papercrete reveals very little existing research and further studies are required to explore further potential [12]. The literature analysis indicates that the amount of paper significantly affects the mechanical properties, dimensional durability and structural integrity of papercrete [13]. The analysis also reveals that the inclusion of residual paper mill waste including waste paper sludge and ash, greatly gives the necessary strength and longevity to the bricks [9]. In most of the studies, the industrial and natural waste such as fly-ash, rice husk ash, and blast furnace slag has been used to improve the strength and other properties, including durability, usability, water absorption capacity, etc [5].

Fly ash is a combustion by-product of coal in thermal stations, removed as fine particle residue from gases before being released into the atmosphere [14]. The tests on fly ash-based bricks demonstrate their suitability to meet the requirements for traditional clay brick specified in ASTM C 216. The cemented brick in the ASTM C 55 and the standard specification for papercrete structural brick also fall within the acceptable shrinkage limits. The process used in fly ash-based bricks is estimated to reduce almost 90% of the construction energy [6]. RHA is another excellent pozzolan used in construction materials. Adding RHA to the concrete mix even in low replacement will dramatically increase the workability, strength and impermeability of concrete mixes, and making the concrete durable to chemical attacks and abrasion [15]. The aim of the present analysis is to manufacture new type of environmentally friendly papercrete bricks by the use of waste materials such as paper, fly ash and rice husk ash to study the strength and durability of papercrete bricks for commercial use in construction works. An experimental program was conducted to investigate the strength and other technological properties of fly ash based papercrete bricks such as dry density, water absorption percentage and tensile strength.

2. Materials and Methods

The main materials for the papercrete are paper and cement. Waste papers from nearby places were collected, and any foreign materials such as pins or threads etc. were removed. Table 1 provides the physical properties of paper. In this study, the entire work was made of 43 grade Ordinary Portland Cement. In addition, fly ash and/or rice husk ash was added in presence of sand. In this study, fly ash was collected from nearby thermal power plant while rice husk was procured from the local supplier and converted to ash. The chemical and physical properties of cement, fly ash and rice husk ash are given in Table 2 and Table 3 respectively. The natural sand used in the study confirmed to zone- II. After chopped in small pieces, the paper pulp was obtained by pulverizing the chopped pieces after dipping in water for 15 days. Excess water was removed by filtration under pressure. The details of the mixed proportions used during the study are given in Table 4. The dry components were sprinkled and uniformly mixed manually over the required amount of paper pulp and then transferred to the mechanical mixer. The wooden mold lined with machine oil on the inner surface were used. Papercrete fibrous mix were poured into the mold in three layers and were hydraulically compressed to obtain bricks with 230 mm x 110 mm x 70 mm in thickness. The specimen was tested for the metallic sound on the brick surface after 15 days and coated externally with waterproofing admixture after having been dried in the air. The compressive strength and split tensile strength were examined according to IS:3495-1992 recommendation. Test for percentage of water absorption and dry density was performed according to ASTM C-67 standards.

Table 1. Physical properties of Paper

Properties	Values
Weight	47 GSM
Thickness	0.06 mm
Moisture	7.5%
Bursting Strength	168 kPa
Tearing resistance	12.6 kg
Tensile Strength	1.13 kg
Porosity	475 mls/minute

Table 2. Chemical Composition of Binder Materials

Oxides % by mass	OPC % by mass	FA % by mass	RHA % by mass
SiO_2	19.46	61.65	91.81
Al_2O_3	4.22	25.86	2.23
Fe_2O_3	3.56	9.56	1.95
CaO	65.92	13.78	0.97
MgO	1.08	2.33	1.92
K_2O	0.67	1.57	-

Na_2O	0.21	1.57	-
TiO_2	0.24	-	-
SO_3	2.6	0.62	0.42
Cl	0.01	-	-
LOI	0.96	2.38	0.9

Table 3. Physical Properties of Components

Aggregate Type	Specific surface area (m^2/kg)	Specific gravity
FA	552	2.40
RHA	187	2.27
Sand	1.15	2.55

Table 4. Mix proportion of Specimens per m^3

Designation	Cement (g)	FA (g)	RHA (g)	Sand (g)	Paper (g)
CM	1200	-	-	3600	0
PC1	960	-	10	3600	120
PC2	900	-	15	3600	120
PC3	840	-	20	3600	120
PC4	780	-	25	3600	120
PC5	960	10	-	3600	120
PC6	900	15	-	3600	120
PC7	840	20	-	3600	120
PC8	780	25	-	3600	120

3. Results and Discussions

3.1. Flow

For all papercrete blends, the flow results are shown in Figure 1. The RHA and FA presence decreased the flow rates as compared to the reference mix. The percentage decreased for flow was observed as 10%, 17%, 23%, 30% respectively at 10%, 15%, 20% and 25% of RHA. Paper fibres can retain water being composed of hydrophilic cellulose. This results in an increased thickness and compact matrix of the papercrete [2]. Addition of RHA has been reported to result in a decline in the workability of the concrete [15]. Results also demonstrate that with the RHA rise, the flow rate declined continuously for all the specimens as compared to the reference mixture [16]. Hence, the combined effect of RHA and paper pulp result in an overall decrease in the flow of the papercrete.

With FA at 10%, 15%, 20% and 25% substitution levels, the reductions in flow were 7%, 7%, 14%, 18% and 24%, respectively. This drop in flow levels is compatible with results reported in literature indicating that although FA has been reported to improve the workability of the concrete, yet the presence of waste paper fibres reduce the slump because the extra water absorbed by the paper fibres compensate the effect [14]. The literature reports that paper fibres absorb water more quickly and effectively than ordinary Portland cement particles resulting in a significant decrease in flow [2]. Moreover, as shown in Figure 1, the FA blends show comparatively more workability than the RHA blends. The finesse of FA in comparison with RHA particles results in better enmeshed structure that can attribute to this behavior [17].

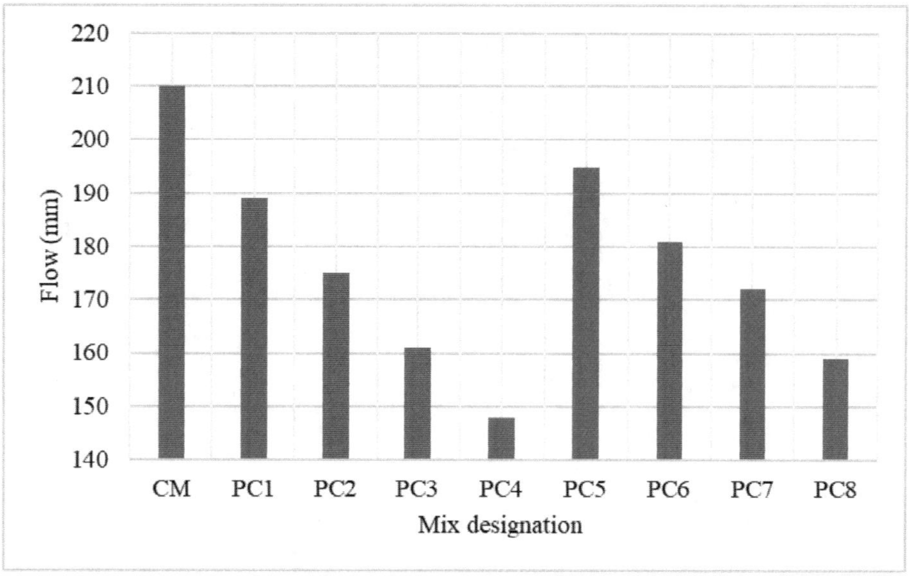

Figure 1. Flow of Specimens

3.2. Compression Strength

Figure 2 demonstrates the compressive strength measurements of the 7-and 28-day specimens. For RHA mixtures, it has been shown that the compressive strength decreased by 24% and 35% at 7 and 28 days respectively in comparison to the reference mixture, with cement being substituted by 10% RHA in presence of 10% paper pulp. This decrease can be due to the role of paper fiber waste which degrades compressive strength due to poor cohesive forces between CSH gel and cellulose particles and decreased amount of cementitious material [8]. By comparison, substitute dosage greater than 10% (15%, 20% and 25%) the compressive strength declined by 16%, 9%, and 13%, respectively and by 19%, 8% and 13% at 7 and 28 days, relative to the reference mix. Literature reports that inclusion of RHA results in enhancement of compressive strength that compensates the negative effect of paper pulp addition and comparatively lesser decrease is obtained in the strength [16]. However, the slight decrease in compressive strength at 25% RHA can be attributed to the increase in friction among the particles at high dosage of RSH [15].

Results revealed that the substitution of 10% of cement with FA provided better results as compared to the RHA blends. Such results have been in line with the literature [17]. At a substitution level of 10%, the strength for FA blends, decreased by 22% and 19%, respectively, at 7 and 28 days. Literature also reports similar results [4]. This degradation can be explained by the loss of cemented activity of the FA in presence of paper due to loss of proper binding. In comparison, at substitution levels of 15% to 20%, the compressive strength levels have been reduced by lesser amount. The reduction rates at 7 and 28 days were 3% and 2% (for a replacement of 15%) and 9% and 6% (for a replacement of 20%). The findings of this analysis are consistent with the pozzolanic effect of FA that produces the strength enhancing CSH gel occupying the cement matrix and compensates the decrease in compressive strength contributed by the low level of silica in paper pulp [14]. However, as the substitution amount of FA was enhanced further to 25%, a slight decrease in compressive strength was observed that may be attributed to the better pozzolanic action of FA at an optimum content and after that satisfactory results are not obtained [18].

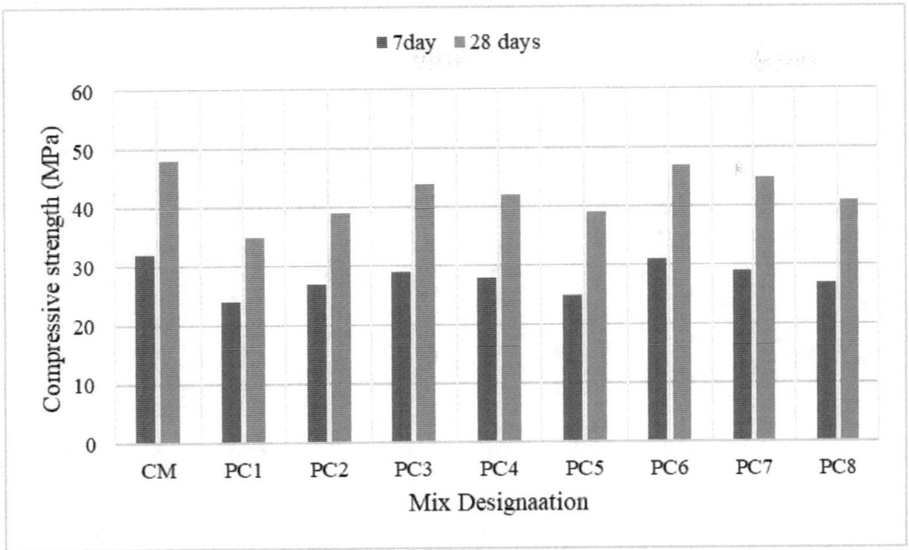

Figure 2. Compressive Strength of Specimens

3.3. Split Tensile Strength

Figure 3 shows the effects of the RHA and FA incorporation on the split tensile strength of specimens at 7 and 28 days. For RHA, the substitution of cement with 10% RHA resulted in a degradation in the tensile strength of the cement mortar in consistence with the compressive strength analysis. The results are coincidental to that of the earlier studies reporting that the tensile strength decreased with introduction of paper pulp. The effect has been assigned to the weak binding between cellulosic fibers and CSH gel [9]. Compared with the reference mix, the reduction was 20% and 16% at 7 and 28 days in presence of 10% paper pulp. Increased cement replacement ratios with RHA to 15%, 20%, and 25% resulted in 14%, 10%, 12%, and 14%, 10%, and 11% decrease in tensile strength at 7 and 28 days, respectively, as compared to the reference mix. Addition of cellulose fibres from the paper pulp has been reported to result in a decreased split tensile strength as compared to reference mix [8]. However, the addition of RHA introduces the fibres that counterattacks the strength deterioration effect of cellulose fibres. This effect is aided by the fine microparticles of RHA that strengthens the binding between the binder particles and regulates the matrix cracking tendency [15]. Fiber intervention will work in two ways: first by space filling, where it is properly designed to avoid the breakage of microcracks in large amounts, and second, by "bridging" the microcracks, which increases the bonding between the two sides of a crack and therefore increases the tensile strength [19].

The 10% cement substitution ratio for FA produced comparable results at 7 and 28 days to that of RHA blends. In addition, there was a decrease of 3%, 6%, 8% at 7 days and 3%, 6%, 9% at 28 days for 15%, 20% and 25% FA substitution. A similar behavior has been found in past studies with pozzolanic activity of FA particles [14]. The presence of fine particles fills the voids and the production of CSH gel increases the bonding, thereby increasing the density and the tensile strength. Further, as the number of voids in the matrix reduces, it increases the tensile strength [18]. However, best results were obtained at 20% RHA and 15% FA in presence of 10% paper pulp that can be amounted as the optimal content of the supplementary cementitious materials.

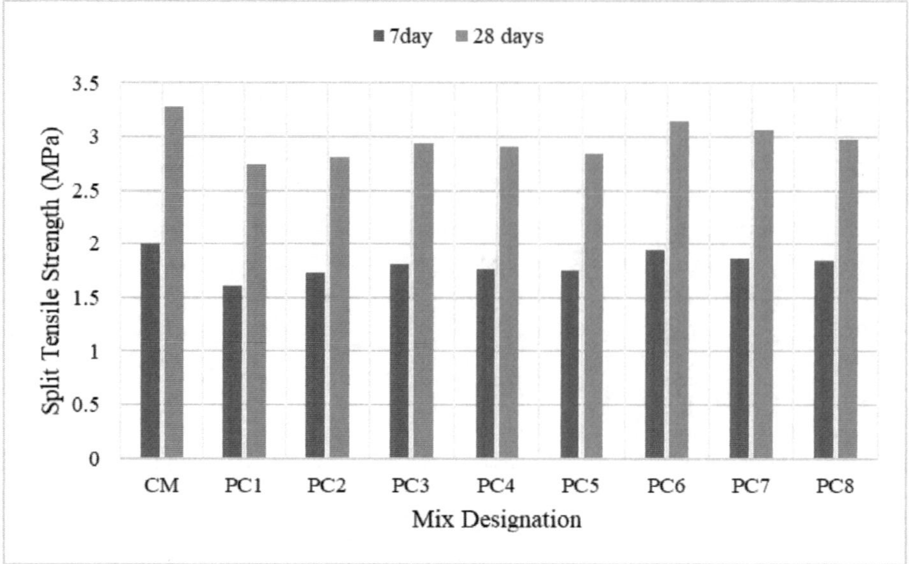

Figure 3. Tensile Strength of Specimens

3.4. Water Absorption %

Figure 4 shows the effects of RHA and FA in presence of 10% paper pulp on water absorption % at 28 days. For RHA, tests have shown that these mixtures were having significantly lower water of absorption than the reference mix but for the substitution dosage of 10% RHA, the results were similar. The presence of cellulose fibres from paper has been reported to decrease the water absorption [20]. Further, higher the level of substitution by RHA, lower the water absorption % was observed. At the replacement dosage of 20% RHA with 9% lower value than the reference mix, the maximum decrease was reported. The large number of voids got occupied in the presence of high quantities of fiber that consequently lead towards a increased density leading to decrease in absorption of water [21]. Studies revealed that water absorption was not improved even in the case of FA mixtures. Water absorption decreased by 5% for the 10% FA dosage, compared to the reference mix. However, the water absorption of the specimens decreased more with the increase in the replacement dosage by 15%. These results are consistent with past studies and have been attributed to the pore filling effect and pozzolanic action of FA [22].

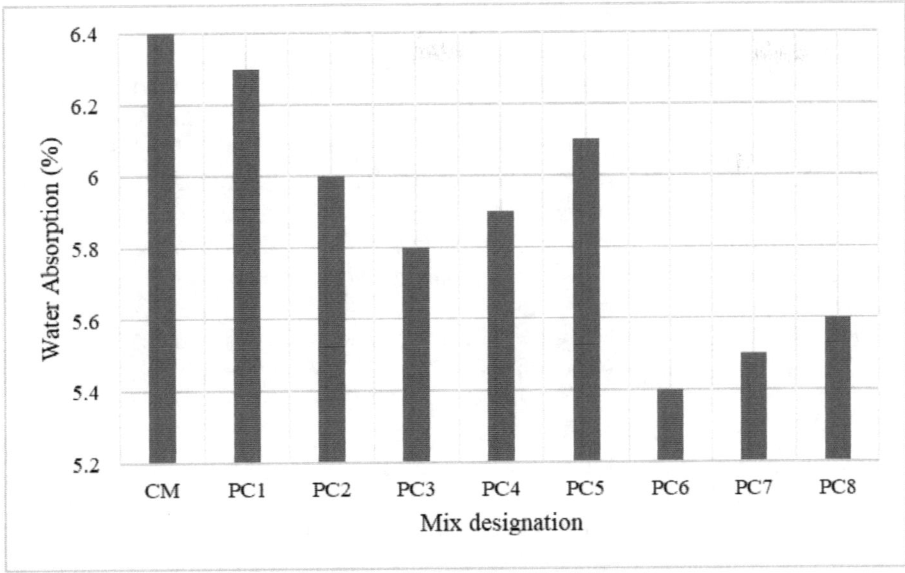

Figure 4. Water Absorption % of Specimens

3.5. Dry Density

Figure 5 illustrates the results of dry density tests for all blends at 28 days. For RHA, the findings showed that the 10% substitution resulted in approximately 1.2% increase in density relative to the reference mix. This increase may be due to the fibrous properties which enable water to be absorbed and retained in the sample during mixing and thus boost the sample density [3]. In comparison with the reference mix, substitution levels above 10% resulted in density increase of 4%, 11% and 8%, respectively, for the replacement content of 15%, 20% and 25%. This is because increased microparticles of RHA occupy the voids in the matrix which increases the density [16]. The results on FA blends have demonstrated that presence of FA particles in the mix increased the density of the papercrete. The increase obtained was 6%, 9%, 7% and 7%, respectively, compared with the reference mix for 10%, 15%, 20% and 25% substitution content of FA. These results coincide with that reported in the literature with increase in density of the matrix by replacement of air voids by microparticles of FA [17].

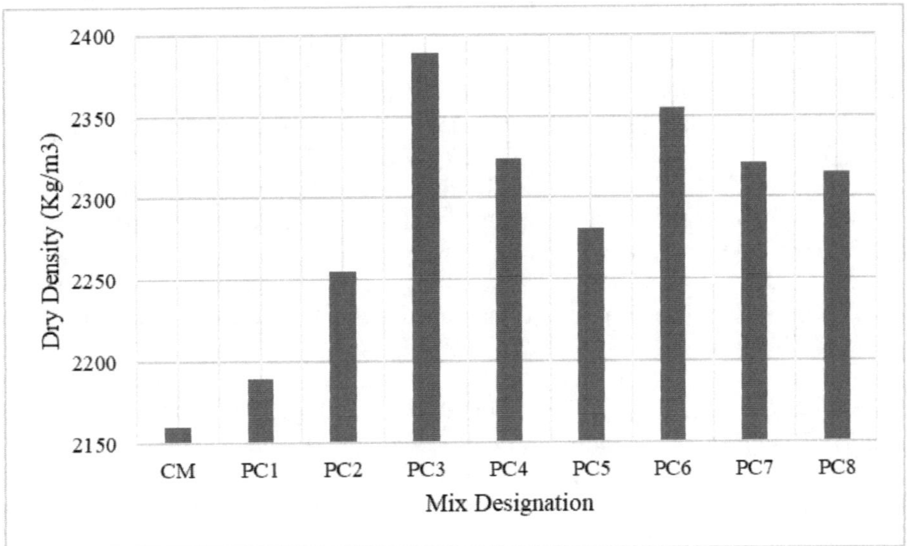

Figure 5. Dry Density of Specimens

4. Conclusion

Based on the study, the following conclusions can be drawn:

1. Replacing cement with RHA and FA in presence of 10% paper pulp has a detrimental effect on the flow of the papercrete. With the rise in the replacement percentages, the flow of all mixes decreased relative to the reference mix.
2. As compared to the reference mix, the compressive and split tensile strength was found lesser for all the mixes at 7 and 28 days of curing. However, among the RHA mixes, maximum compressive and split tensile strength was obtained for 20% RHA in presence of 10% paper pulp. On the other hand, among the FA mixes, maximum compressive and split tensile strength was obtained for 15% FA in presence of 10% paper pulp.
3. The substitution of cement by 10% paper pulp in presence of RHA and FA decreased the water absorption of all the mixes as compared to the reference mix. However, minimum water of absorption was obtained at 15% FA in presence of 10% paper pulp followed by 20% RHA in presence of 10% paper pulp.
5. Maximum dry density was obtained for the strengthened samples.

Hence, 15% FA in presence of 10% paper pulp and 20% RHA in presence of 10% paper pulp can be stated as the optimal content for improved properties of papercrete that can be used for practical applications.

References

[1] Stevulova N, Schwarzova I, Hospodarova V and Junak J 2016 Implementation of waste cellulosic fibres into building materials *Chem. Eng. Trans.* **50** 367–72
[2] Hospodarova V, Stevulova N, Briancin J and Kostelanska K 2018 Investigation of Waste Paper Cellulosic Fibers Utilization into Cement Based Building Materials *Buildings* **8** 43
[3] Chen M, Zheng Y, Zhou X, Li L, Wang S, Zhao P, Lu L and Cheng X 2019 Recycling of paper sludge powder for achieving sustainable and energy-saving building materials *Constr. Build. Mater.* **229** 116874
[4] Shibib K S 2015 Effects of waste paper usage on thermal and mechanical properties of fired brick *Heat Mass Transf. und Stoffuebertragung* **51** 685–90
[5] Muñoz P, Letelier V, Muñoz L and Bustamante M A 2020 Adobe bricks reinforced with paper & pulp wastes improving thermal and mechanical properties *Constr. Build. Mater.* **254**
[6] Vashistha P, Kumar V, Singh S K, Dutt D, Tomar G and Yadav P 2019 Valorization of paper mill lime sludge via application in building construction materials: A review *Constr. Build. Mater.* **211** 371–82
[7] Yun H, Jung H and Choi C 2011 Mechanical properties of papercrete containing waste paper *ICCM International Conferences on Composite Materials* pp 1–4
[8] Zaki H, Gorgis I and Salih S 2018 Mechanical properties of papercrete ed T S Al-Attar, M A Al-Neami and W S AbdulSahib *MATEC Web Conf.* **162** 02016
[9] Kubba H Z, Nasr M S, Al-Abdaly N M, Dhahir M K and Najim W N 2020 Influence of Incinerated and Non-Incinerated waste paper on Properties of Cement Mortar *IOP Conf. Ser. Mater. Sci. Eng.* **671** 012113
[10] Malaiskiene J, Kizinievic O, Kizinievic V and Boris R 2018 The impact of primary sludge from paper industry on the properties of hardened cement paste and mortar *Constr. Build. Mater.* **172** 553–61
[11] bin Mohd Sani M S H, bt Muftah F and Rahman M A 2011 Properties of Waste Paper Sludge Ash (WPSA) as cement replacement in mortar to support green technology material *2011 3rd International Symposium & Exhibition in Sustainable Energy & Environment (ISESEE)* (IEEE) pp 94–9
[12] Ahmadi B and Al-Khaja W 2001 Utilization of paper waste sludge in the building construction industry *Resour. Conserv. Recycl.* **32** 105–13
[13] Dalmay P, Smith A, Chotard T, Sahay-Turner P, Gloaguen V and Krausz P 2010 Properties of

cellulosic fibre reinforced plaster: influence of hemp or flax fibres on the properties of set gypsum *J. Mater. Sci.* **45** 793–803

[14] Bui N K, Satomi T and Takahashi H 2019 Influence of industrial by-products and waste paper sludge ash on properties of recycled aggregate concrete *J. Clean. Prod.* **214** 403–18

[15] Molaei Raisi E, Vaseghi Amiri J and Davoodi M R 2018 Mechanical performance of self-compacting concrete incorporating rice husk ash *Constr. Build. Mater.* **177** 148–57

[16] Chopra D, Siddique R and Kunal 2015 Strength, permeability and microstructure of self-compacting concrete containing rice husk ash *Biosyst. Eng.* **130** 72–80

[17] Rukzon S and Chindaprasirt P 2009 An Experimental Investigation of the Carbonation of Blended Portland Cement Palm Oil Fuel Ash Mortar in an Indoor Environment *Indoor Built Environ.* **18** 313–8

[18] Kearsley E P and Wainwright P 2001 the Effect of fly ash content on the compressive strength development of concrete *Cem. Concr. Res.* **31** 105–12

[19] Fapohunda C, Akinbile B and Shittu A 2017 Structure and properties of mortar and concrete with rice husk ash as partial replacement of ordinary Portland cement – A review *Int. J. Sustain. Built Environ.* **6** 675–92

[20] Gomes L M, Aroche A, Schafer M, Erhart R, Moraes C A M, de Campos Rocha T L A and Brehm F A 2014 Influence of Cellulose Pulp Waste in Plastering Mortar *Key Eng. Mater.* **634** 222–34

[21] Siddique R, Singh K, Kunal P, Singh M, Corinaldesi V and Rajor A 2016 Properties of bacterial rice husk ash concrete *Constr. Build. Mater.* **121** 112–9

[22] Mardani-Aghabaglou A, Inan Sezer G and Ramyar K 2014 Comparison of fly ash, silica fume and metakaolin from mechanical properties and durability performance of mortar mixtures view point *Constr. Build. Mater.* **70** 17–25

IOP Conf. Series: Materials Science and Engineering **961** (2020) 012011 doi:10.1088/1757-899X/961/1/012011

Green Synthesis and Characterization of Copper Oxide Nanoparticles Using *Psidium guajava* Leaf Extract

Ammu Varughese[1], Raminder Kaur[1] & Poonam Singh[1]

[1]Department of Applied Chemistry, Delhi Technological University, Delhi 110042, India

ammuvarughese319@gmail.com, raminderkaur@dce.ac.in, poonam@dtu.ac.in

Abstract Green synthesis of nanoparticles has gained enormous attention in today's world due to ongoing demand to develop safe, sustainable, cost-effective and environmentally friendly process for synthesizing a wide variety of materials including metal/metal oxides nanoparticles, hybrid and bioinspired materials. In the current study, we have carried out green synthesis of copper oxide (CuO) nanoparticles (NPs) using *psidium guajava* leaf extract as capping agent and copper acetate as metal precursor. The biosynthesized nanoparticles were characterized using wide variety of techniques i.e. powder X-ray diffraction (PXRD), FTIR, TGA and HRTEM. The PXRD result confirms the synthesis of copper oxide nanoparticles in pure phase having monoclinic symmetry. The average particle size using PXRD was found out to be ~33 nm.

Key words: *ecofriendly, cost-effective, sustainable, nanoparticles and metal oxides*

1. Introduction

Over the last few decades, synthesis of nanomaterials (such as metal NPs, carbon nanotubes (CNTs), quantum dots (QDs), graphene, etc.) having dimension within the range of 10-100 nm have become an extensive area of research due to their potential applications in broad area of science and technology [1-6]. Though bulk synthesis of nanoparticles can be carried out rapidly using conventional methods (i.e. physical and chemical methods), they require use of hazardous chemicals [7-9]. Therefore keeping this in mind, use of green synthesis approach that makes use of mild reaction conditions and non-hazardous precursors have been emphasized for promoting environmental sustainability [10-15]. The biological systems that are actively used in the green synthesis of NPs include plants and their derivatives (phytosynthesis), microorganisms like fungi, bacteria, algae and yeast [16-17]. Among all the available methods, phytosynthesis is the ideal method for the synthesis of metal oxide NPs owing to several benefits associated with it i.e. safe, efficient, cost-effective, simple (does not involve complex & multistage process like microbial isolation, culturing, etc.) and provides better control over shape, size & dispersion of metal nanoparticles in comparison to bacteria and/or fungi mediated synthesis [18].

Among a variety of transition metal oxides, CuO is one amongst the potential *p*-type semiconductors that have gained considerable attention attributed to its excellent physical, electrical, optical, and magnetic properties. CuO with a narrow band gap of 1.2 eV is widely used in catalysis [19-22], electrochemistry [23-24], solar energy conversion [25], sensors/biosensors [26-28], energy storage [29], biocidal agents [30-31] etc. In literature, green synthesis of CuO NPs is reported using plants such as *Calotropis procera* leaf extract [32], *Abutilon indicum* leaf extract [33], aloe vera leaf extract [34], karaya gum [35], Cassia alata flower extract etc.

In the current study, *Psidium guajava* leaves have been used for carrying out the synthesis of copper oxide NPs due to the presence of high amount of phytochemicals in it [36-37]. *Psidium guajava* is a medicinal plant commonly known as guava & has a place in Myrtle family as Myrtaceae [38]. *Psidium guajava* is found in various areas of the tropical and subtropical zones and has been reported to have antidiarrheal [39], antispasmodic [40], antioxidant [41], anti-allergy [42], antibacterial [43], anti-cough [44], anti-inflammatory [45], and anticancer activities [46].

2. Materials and methods

Copper acetate monohydrate (Merck, $\geq$ 98.0 %), Sodium lauryl sulfate (Sisco Research Laboratories $\geq$ 98.0 %) and fresh *psidium guajava* leaves collected from the Delhi Technological University (DTU) campus.

IOP Conf. Series: Materials Science and Engineering **961** (2020) 012011 doi:10.1088/1757-899X/961/1/012011

2.1. Extraction of psidium guajava Leaf Extract:

Fresh *psidium guajava* leaves collected from the DTU campus were first cut into small pieces. These leaves were then washed thoroughly using double distilled water to eliminate debris. These cleaned leaves were then dried in the presence of sunlight for 5 days. After that the dried leaves were ground into a fine powder using pestle mortar & stored in dry place.

Psidium guajava leaf extract was prepared by heating the solution of *psidium guajava* leaf powder and distilled water (1:10 wt. %) at 100°C for 30 minutes in Erlenmeyer flask till the colour of solution changes to dark brown (indicating phytochemical extraction). After that the solution was allowed to cool down to room temperature and filtered through Whatman No.1 filter paper. The filtrate was used for the synthesis of copper oxide NPs.

2.2. Synthesis of CuO Nanoparticles:

To synthesise CuO NPs 1 M copper acetate monohydrate was dissolved in 10 ml of distilled water at room temperature using magnetic stirrer (250 rpm). After that, 30 ml of *psidium guajava* leaf extract was added into the copper acetate solution drop wise, with continuous stirring. Later the solution was kept on magnetic stirrer (250 rpm) for 24 hrs at 60 °C leading to the formation of gel that was dried in hot air oven at temperature of 60 °C. The green precipitate thus obtained was calcined at temperatures of 400 °C for 6 hrs in the muffle furnace, forming black CuO nanoparticles.

2.3. Synthesis of CuO Nanoparticles in presence of sodium lauryl sulfate (SLS)

For the synthesis of CuO nanoparticles in presence of SLS, SLS (4 g) was added into p*sidium guajava* leaf extract (30 ml) and the resulting solution was stirred for five minutes. Rest of the procedure followed was same as mentioned above.

PXRD pattern of synthesized sample was recorded using high-resolution Bruker D8 Advanced X Ray diffractrometer employing Cu Kα radiation ($\lambda = 1.5418$ Å) over the range of $2\theta = 10\text{-}80°$. Thermo gravimetric analysis (TGA) was conducted using the Pyris 1® TGA in the range 50-800°C under flowing nitrogen (50 mL min^{-1}) at a heating rate of 10 °C min^{-1}. FTIR spectra of the sample was recorded using a Perkin-Elmer 2000 FTIR spectrometer

IOP Conf. Series: Materials Science and Engineering **961** (2020) 012011 doi:10.1088/1757-899X/961/1/012011

employing KBr disks. TEM of the sample was recorded using Jeol/JEM 2100 HRTEM operated at an acceleration voltage of 200KV.

3. Results and discussion

3.1. Qualitative Phytochemical Screening:

Phytochemical screening of freshly prepared *psidium guajava* leaf extract was carried out using simple chemical tests to identify the presence of active phytoconstituent i.e. polyphenol, alkanoids, flavonoids, saponins, tannins, etc. in the sample (Table 1).

Table 1 Phytochemical screening of *psidium guajava* leaf extract

Component	Procedure	Result
Flavonoids	A portion of aqueous extract was added to 5 ml of aq ammonia solution, followed by the addition of Conc. H_2SO_4 [47].	Absence of yellow colour indicating the absence of flavonoids.
Gum & Mucilages	10 ml of extract was added to 25 ml of ethanol with constant stirring [48].	No white or cloudy precipitate indicating the absence of Gum & Mucilages
Phenolic Compounds	To 5 ml of leaf extract few drops of 5% neutral $FeCl_3$ was added [49].	formation of dark green colour indicating the presence of Phenolic Compounds
Terpenoids	To 5 ml of leaf extract 2 ml of chloroform was added followed by the addition of 2 ml of Conc. H_2SO_4 carefully along the sides of the test tube to form a layer [50].	No formation of a reddish brown colour at the interface indicates the absence of terpenoids.
Saponins	A 1 ml aliquot of the leaf extract was diluted using 20 ml of distilled water and was then shaken vigorously for 15 minutes in a vortex [49].	Persistent foaming indicates the presence of saponins.
Glycosides	Few drops of aqueous NaOH was added into a 1ml of leaf extract [47].	No formation of yellow color indicates the absence of Glycosides.

Steroids	5 ml of extract was added to a 2 ml of chloroform & Conc. H_2SO_4 solution and was shaken thoroughly [51].	No formation of a reddish brown colour at the lower chloroform layer indicates the absence of steroids.

Phytoconstituent are involved in the green synthesis of metal oxide nanoparticles and can acts as both effective reducing as well as stabilizing agents. Therefore in order to understand the mechanism of nanoparticles synthesis screening of phytochemical is inevitable. Phytochemical analysis exhibit positive result for phenolic & saponins content in the leaf extract inferring its significant properties and shows negative results for flavonoids, gum & mucilages, terpenoids, glycosides and steroids.

3.2. Structural Analysis of CuO Nanoparticles

The PXRD pattern of CuO nanoparticles synthesized using *psidium guajava* leaf extract in absence and in presence of SLS is presented in Figure 1. The presence of PXRD peaks at 2θ value of 32.53°, 35.63°, 38.74°, 48.83°, 53.56°, 58.36°, 61.51°, 65.85°, 66.42°, 68.11°, 72.34° and 75.33° confirms the formation of CuO having monoclinic symmetry with lattice constant a = 4.683, b = 3.428, c = 5.129 Å (JCPDS 80-1268) [52]. Since no additional peak due to the presence of any other phase was observed in absence of SLS this suggests the formation of copper oxide nanoparticles in pure phase.

The average particle size of synthesized copper oxide nanoparticles was calculated using Debye- Scherrer's formula, i.e.,

$$\mathbf{D} = \frac{\mathbf{k\lambda}}{\boldsymbol{\beta}\mathbf{Cos\theta}}$$

Where, D, λ, β, θ represents average particle size (nm), wavelength of x-ray (0.15406 nm), full width half maximum of the intense peak, Bragg's angle and k is taken as 0.89 (constant), respectively. Using the above equation, the average particle size of copper oxide nanoparticles was found out to be ~33 nm.

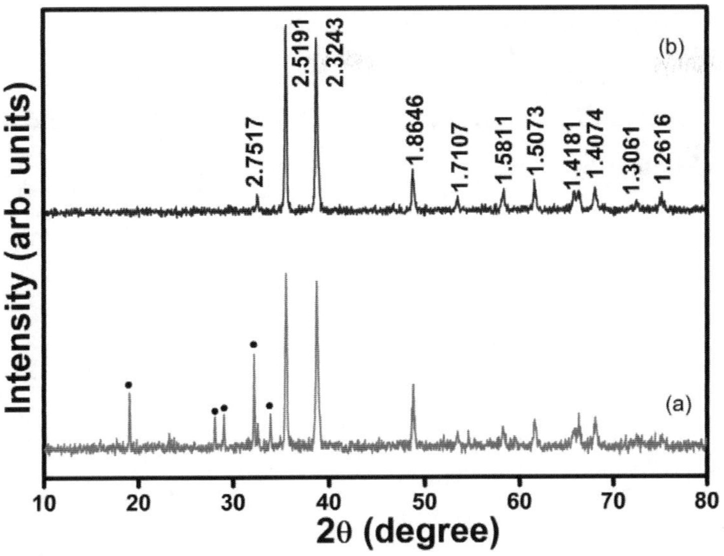

Figure 1: PXRD pattern of CuO nanoparticles in presence (a) and in absence of SLS (b). "•" denote reflections due to Na_2SO_4

Further confirmation of internal structure, accurate measurement of particle size and morphology was done using TEM analysis (Figure: 2). Both the average particle size & interplanar spacing values were in close agreement to that obtained from XRD data. However SAED pattern confirms the crystalline nature of synthesized copper oxide nanoparticles.

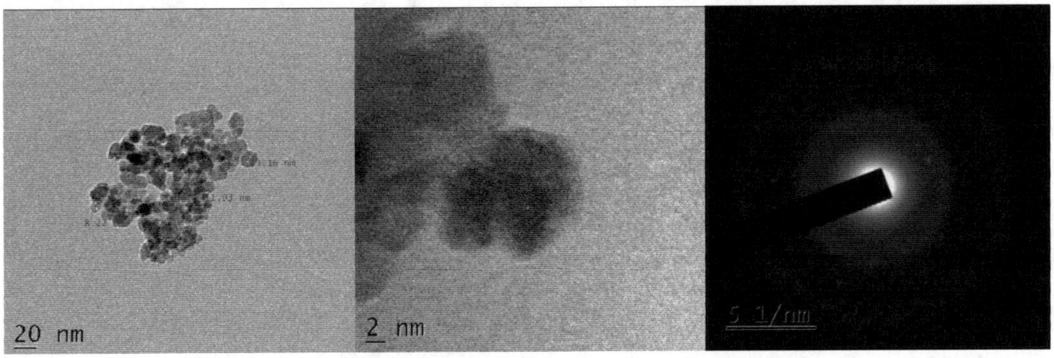

Figure 2: HR-TEM image and SAED pattern of CuO nanoparticles

The FTIR spectrum of synthesized copper oxide nanoparticles is presented in Figure 3. The peaks around 784, 624 and 529 cm^{-1} corresponds to the Cu–O stretching vibration of copper oxide nanoparticles in the monoclinic structure. The absorption peaks at 3439 cm^{-1} and 1628 cm^{-1} corresponds to the OH stretching vibration and HOH bending mode of adsorbed water molecules, since the nano crystalline materials possess high surface area to volume ratio leading to the absorption of moisture in the lattice. The absorption band at 1107cm^{-1} corresponds to the C–O stretching of phenol and alcoholic compounds.

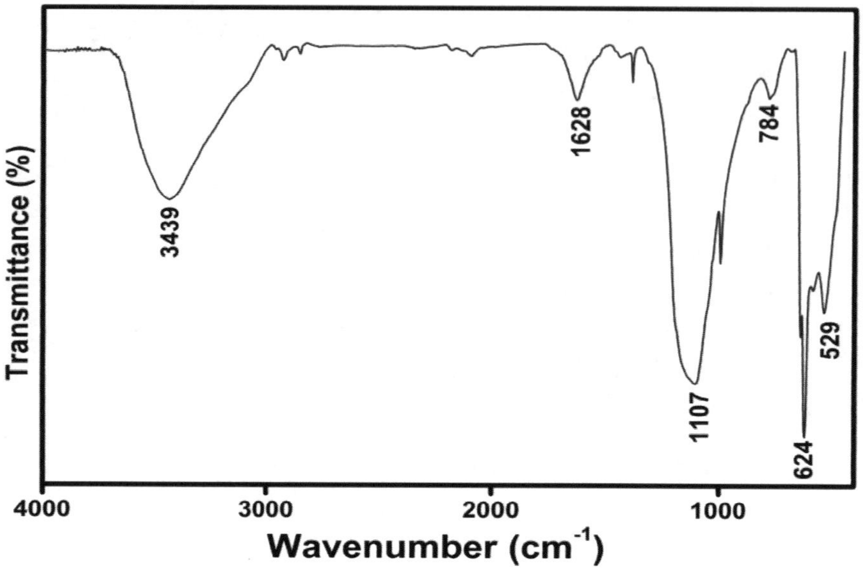

Figure 3: FTIR spectrum of CuO nanoparticles

The TGA/DTG profile of copper oxide nanoparticles is presented in Figure 4. The TGA curve showed a three-step decomposition process of the precursor to form CuO NPs. The first weight loss below 166°C (~9%) corresponds to the removal of moisture and organic solvent molecules. The second and third weight loss in the region from 166 – 403°C may be due to pyrolysis and combustion of organic compounds form precursor and solvent that leaves the system in the form of CO, CO_2 and other organic gases. At further higher temperatures, no significant weight loss was observed, thereby supporting the formation of copper oxide NPs with high purity.

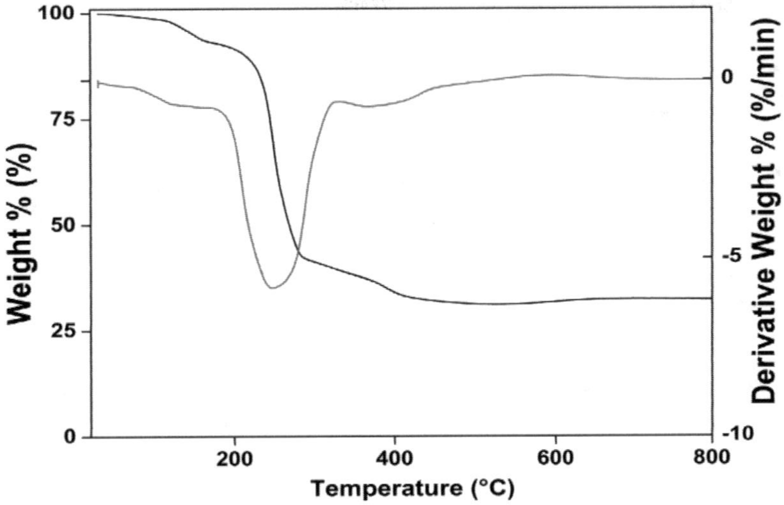

Figure 4: TGA/DTG thermograph of CuO precursor (under N_2 atmosphere)

Hence, based on TGA and phytochemical analysis, the mechanism of biochemical reduction & stabilization of CuO nanoparticles might be as follows:

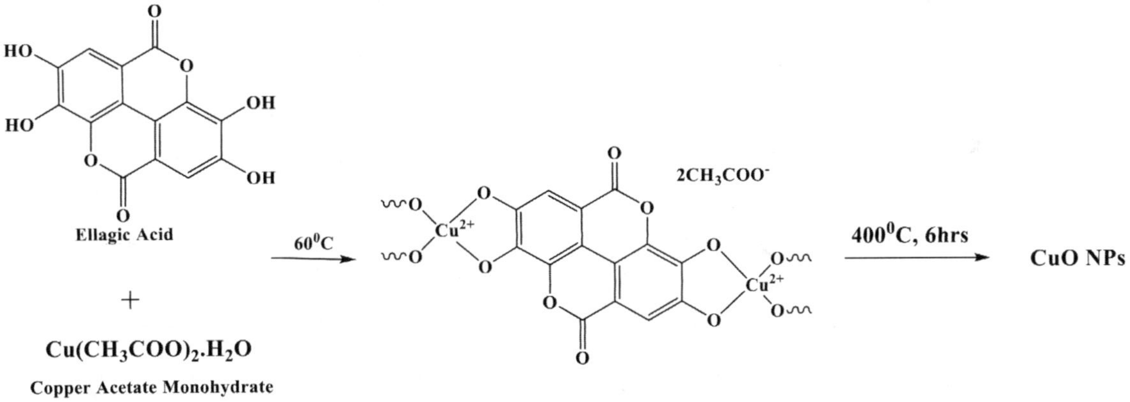

(**Figure 5:** Mechanism of the synthesis of CuO nanoparticles)

4. Conclusion

The present study reports the biosynthesis of copper oxide nanoparticles by making use of using *psidium guajava* leaf extract as both capping & reducing agent. The presence of phytoconstituent (phenolic compounds) in the leaf extract plays major role in the biochemical reduction & stabilization of copper oxide at nano scale level. The PXRD spectrum confirms the formation of CuO in monoclinic phase having average crystallite size of ~33nm. The HRTEM revealed spherical morphology of CuO nanoparticles. The methodology used for the synthesis is green and viable because of its ease, efficiency, cost-efficient and eco-friendly nature in comparison to other methods.

5. References

1) Y. Yang, C. Zhang & Z. Hu, **2013**, "Impact of metallic and metal oxide nanoparticles on waste water treatment and anaerobic digestion", *Environmental Science: Processes & Impacts*, **15**, 39.
2) T.W.Woolerton, S.Sheard, E.Pierce, S.W.Ragsdaleb & F.A.Armstrong, 2011, "CO_2 photoreduction at enzyme-modified metal oxide nanoparticles", Energy Environ. Sci., **4**, 2393–2399.
3) S.A. Corr, 2013, "Metal Oxide Nanoparticles", *Nanoscience*, **1**, 180–207.
4) R.Fang, P.Tian, X.Yang, R.Luque, Y.Li, 2012, "Encapsulation of Ultrafine Metal-Oxides Nanoparticles within Mesopores for Biomass-Derived Catalytic Applications", *J. Chem, Science*, **00**, 1-3.
5) Z.Zhang, X.Zou, L.Xu, L.Liao, W.Liu, J.Ho, X.Xiao, C.Jiang, J.Li, 2015, "Hydrogen gas sensor based on metal oxide nanoparticles decorated graphene transistor", *Nanoscale*.
6) N. Verma, N. Kumar, "Synthesis & Biomedical Application of Copper Oxide Nanoparticles: An Expanding Horizon, 2019, *ACS Biomater. Sci. Eng*. **5**, 1170-1188.
7) K.B. Narayanan, N. Sakthivel, "Biological synthesis of metal nanoparticles by microbes", 2010, *Elsevier Advances in Colloid and Interface Science,* **156**, 1–13.
8) P. Dauthal, M. Mukhopadhyay, 2016, "Nobel Metal Nanoparticles: Plant-mediated Synthesis, Mechanistic Aspects of Synthesis, and Applications", *Ind. Eng. Chem. Res*, **55**, 9557-9577.
9) C.P. Devatha, A.K. Thalla, "Green Synthesis of Nanomaterials", *Synthesis of Inorganic Nanomaterials,* 169-184.
10) Anastas P. T, Warner J. C., 1998, Green Chemistry: Theory and Practice; Oxford University Press, Inc.: New York.
11) P. Malik, Ravi Shankar, V. Malik, N Sharma, T. K. Mukherjee, 2014, "Green Chemistry Based Benign Routes for Nanoparticle Synthesis", *Journal of Nanoparticles*, **Volume 2014**, Article ID 302429, 14 pages.

12) J. Virkutyte, R.S. Varma, 2013, "Green Synthesis of Nanomaterials: Environmental Aspects", *ACS Symposisum Series.*
13) P. Raveendran, J. Fu, S.L. Wallen, 2003, "Completely "Green" Synthesis and Stabilization of Metal Nanoparticles", *J. AM. CHEM. SOC.*, **125,** 13940-13941.
14) V. Srivastava, D. Gusain, Y.C. Sharma, 2015, "Critical Review on some widely used Engineered Nanoparticles", *Ind. Eng. Chem. Res,* **54**, 6209-6233.
15) D.L. Hjeresen, D.L. Schutt, J.M. Boese, 2000, "Green Chemistry & Approach", *Journal of Chemical Education,* **Vol 77**, No. 12, 1543-1547.
16) S. Pirtarighat, M. Ghannadnia, S. Baghshahi, 2019, "Green synthesis of silver nanoparticles using the plant extract of Salvia spinosa grown in vitro and their antibacterial activity assessment", *Journal of Nanostructure in Chemistry*, **9**, 1–9.
17) P. Khandel, R.K. Yadaw, D.K. Soni, L. Kanwar, S.K. Shahi, 2018, "Biogenesis of metal nanoparticles and their pharmacological applications: present status and application prospects", *Journal of Nanostructure in Chemistry*, **8**, 217–254.
18) M.S. Akhtar, J. Panwar, YS Yun, 2013, "Biogenic Synthesis of Metallic Nanoparticles by Plant Extracts", *ACS Sustainable Chem. Eng.,* **1**, 591-602.
19) Z. Su, J. Li, D. Zhang, P. Ye, H. Li, Y. Yan, 2019, "Novel flexible Fenton-like catalyst: Unique CuO nanowires arrays on copper mesh with high efficiency across a wide pH range", *Science of the Total Environment*, **647**, 587–596.
20) M.A. Bhosale, B.M. Bhanage, 2014, "A facile one-step approach for the synthesis of uniform spherical Cu/Cu_2O nano- and microparticles with high catalytic activity in the Buchwald–Hartwig amination reaction", *RSC Adv.*, **4**, 15122.
21) S.Momeni, F.Sedaghati, 2018, "CuO/Cu_2O nanoparticles: A simple and green synthesis, characterization and their electrocatalytic performance towards formaldehyde oxidation", *Microchemical Journal,* **143**, 64.
22) X. Liu, H. Jia, Z. Sun, H. Chen, P. Xu, P. Du, 2014, "Nanostructured copper oxide electrodeposited from copper(II)complexes as an active catalyst for electrocatalytic oxygen evolution reaction", *Electrochemistry Communications,* **46**, 1-4.
23) C.W. Li, J. Ciston, M.W. Kanan, 2014, "Electroreduction of carbon monoxide to liquid fuel on oxide-derived nanocrystalline copper", *Nature*, **Vol 000**, 1-4.

24) A. Pendashteh, M.F. Mousavi, M.S. Rahmanifar, 2013, "Fabrication of anchored copper oxide nanoparticles on graphene oxide nanosheets via an electrostatic coprecipitation and its application as Supercapacitor", *Electrochimica Acta*, **88**, 347– 357.

25) H. Siddiqui, M.R. Parra, P. Pandey, M.S. Quereshi, F. Z. Haque, 2020, "Utility of copper oxide nanoparticles (CuO-NPs) as efficient electron donor material in bulk-heterojunction solar cells with enhanced power conversion efficiency", *Journal of Science: Advanced Materials and Devices,* **Vol. 5**, Issue 1, 104-110.

26) A.D. Tocco, S.N. Robledo, Y. Osuna, J. Sandoval-Cortez, A.M. Granero, N.R. Vettorazzi, J.L. Martínez, E.P. Segura, A. Iliná, M.A. Zon, F.J. Arévaloa,, H. Fernández, 2018, "Development of an electrochemical biosensor for the determination of triglycerides in serum samples based on a lipase/magnetite-chitosan/copper oxide nanoparticles/multiwalled carbon nanotubes/pectin composite", *Talanta*, **190**, 30–37.

27) T. Alizadeh, S. Mirzagholipur, 2014, "A Nafion-free nonenzymatic amperometric glucose sensor based on copper oxide nanoparticles-graphene nanocomposite", *Sensors and Actuators B*, **198**, 438-447.

28) F.E. Annanouch, Z. Haddi, S. Vallejos, P. Umek, P. Guttmann, C. Bittencourt, E. Llobet, 2015, "Aerosol assisted CVD grown WO3 nanoneedles decorated with copper oxide nanoparticles for the selective and humidity resilient detection of H2S", ACS Appl. Mater. Interfaces, **7**, 12, 6842-6851.

29) P. Liang, F. Wang, Z. Hu, 2018, "Controlled synthesis of ordered sandwich CuCo2O4/reduced graphene oxide composites via Layer-by-layer hetero-assembly for high-performance supercapacitors", *Chemical Engineering Journal*, **350**, 627.

30) S.M. Dizaj, F. Lotfipour, M.B. Jalali, M.H. Zarrintan, K. Adibkia, 2014, "Antimicrobial activity of the metals and metal oxide nanoparticles", *Materials Science and Engineering*, **44**, 278–284.

31) R. Sivaraj, P.K.S.M. Rahman, P. Rajiv, H.A. Salam, R. Venkatesh, 2014, "Biogenic copper oxide nanoparticles synthesis using Tabernaemontana divaricate leaf extract and its antibacterial activity against urinary tract pathogen", *Spectrochimica Acta Part A: Molecular and Biomolecular Spectroscopy*, **133**, 178–181.

32) K. Rayapa Reddy, 2017, "Green synthesis, morphological and optical studies of CuO nanoparticles", *Journal of Molecular Structure*, **1150**, 553-557.

33) F. Ijaz, S. Shahid, S.A. Khan, W. Ahmad, S. Zaman, 2017, "Green synthesis of copper oxide nanoparticles using Abutilon indicum leaf extract: Antimicrobial, antioxidant and photocatalytic dye degradation activities", *Tropical Journal of Pharmaceutical Research*, **16**, 4: 743-753.
34) P. P. N. Vijay Kumar, U. Shameem, P. Kollu, R. L. Kalyani & S. V. N. Pammi, 2015, "Green Synthesis of Copper Oxide Nanoparticles Using Aloe vera Leaf Extract and Its Antibacterial Activity against Fish Bacterial Pathogens", *BioNanoScience,* **vol. 5**, 135–139.
35) Vinod Vellora Thekkae Padil & Miroslav Černík, 2013, "Green synthesis of copper oxide nanoparticles using gum karaya as a biotemplate and their antibacterial application", *International Journal of Nanomedicine*, **8**, 889–898.
36) E. D-Cerio, AM G-Caravaca, V Verardo, A F-Gutierrez & A S-Carretero, 2016, "Determination of guava (Psidium guajava L.) leaf phenolic compounds using HPLC-DAD-QTOF-MS", *Journal of Functional Foods*, **22**, 376–388.
37) E. D-Cerio, AM G-Caravaca, V Verardo, A F-Gutierrez & A S-Carretero, 2015, "Determination of Polar Compounds in Guava Leaves Infusions and Ultrasound Aqueous Extract by HPLC-ESI-MS", *Journal of Chemistry*, **Volume 2015**, Article ID 250919, 9 pages.
38) T. Santhoshkumar, AA Rahman, C Jayaseelan, G Rajakumar, S Marimuthu, AV Kirthi, K Velayutham, J Thomas, J Venkatesan, SK Kim, 2014, "Green synthesis of titanium dioxide nanoparticles using Psidium guajava extract and its antibacterial and antioxidant properties", *Asian Pacific Journal of Tropical Medicine*, 968-976.
39) Ojewole JA, Awe EO, Chiwororo WD, 2008, "Antidiarrheal activity of Psidium guajava Linn. (Myrtaceae) leaf aqueous extract in rodents", *J Smooth Muscle Res*, **44**, 6, 195-207.
40) X. Lozoya, H. Reyes-Morales, M.A Chávez-Soto, M.C. Martínez-García, Y. Soto-González, S.V. Doubova, 2002, "Intestinal anti-spasmodic effect of a phytodrug of Psidium guajava folia in the treatment of acute diarrheic disease", *J Ethnopharmacol,* **83**, 1-2, 19-24.
41) S.M. El-Amin, M.A.M. Hashash, A.M. Abdou, A.M. Saad, M.S. Abdel-Aziz, A.S. Mohamed, 2016, "Antimicrobial and antioxidant activities of Psidium guajava leaves growing in Egypt", *Der Pharmacia Lettre*, **8**, 12, 27-33.

42) N. Seo, T. Ito, N. Wang, X. Yao, Y. Tokura, F. Furukawa, M. Takigawa, S Kitanaka, 2005, "Anti-allergic Psidium guajava extracts exert an antitumor effect by inhibition of T regulatory cells and resultant augmentation of Th1 cells", *Anticancer Res*, **25**, 6A, 3763-70.
43) Neviton RS, Aparicio DGC, Simone MS, Vataru CN, Prado BF, 2005, "An evaluation of antibacterial activity of Psidium guajava (L)", *Braz Arch Biol Technol*, **8**, 3, 429-436.
44) P. Jaiarj, P. Khoohaswan, Y. Wongkrajang, P. Peungvicha, P. Suriyawong, M.L. Sumal Saraya, O. Ruangsomboon, 1999, "Anticough and antimicrobial activities of Psidium guajava Linn. leaf extract", *Journal of Ethnopharmacology,* **Vol 67**, 2, 203-212.
45) Ojewolw JA, 2006, "Antiinflammatory and analgesic effects of Psidium guajava Linn (Myrtaceae) leaf aqueous extracts in rats and mice", *Meth Find Exp Clin Pharmacol*, **28** 7, 441-446.
46) Sang-Bang L, Hae-Ryong P, 2010, "Anticancer activity of guava (Psidium guajava L.) branch extract against HT-29 human colon cancer cells", *J Med Plant Res*, **4**, 10, 891-896.
47) Emasushan M, John Britto S, 2018, "Preliminary phytochemical profiling and antifungal activity of the seeds and pericarp of Putranjiva roxburghii Wall", *The Pharma Innovation Journal,* **7**, 4, 107-110.
48) C. Uma, K.G. Sekar, 2014, "Phytochemical analysis of a folklore medicinal plant citrullus colocynthis L (bitter apple)", *Journal of Pharmacognosy and Phytochemistry*, **2**, 6, 195-202.
49) K.S. Banu, L. Catherine, 2015, "General Techniques Involved in Phytochemical Analysis", *IJARCS*, **2**, 4, 25-32.
50) R.S. Kumar, C. Venkateshwar, G. Samuel, S.G. Rao, 2013, "Phytochemical Screening of some compounds from plant leaf extracts of Holoptelea integrifolia (Planch.) & Celestrus emarginata (Grah.) used by Gondu tribes at Adilabad District, Andhrapradesh, India", International Journal of Engineering Science Invention, **Volume 2**, Issue 8, PP.65-70.
51) R. Gul, S.U. Jan, S. Faridullah, S. Sherani, N. Jahan, 2017, "Preliminary Phytochemical Screening, Quantitative Analysis of Alkaloids, and Antioxidant Activity of Crude Plant Extracts from Ephedra intermedia Indigenous to Balochistan", *Hindawi Scientific World Journal*, **Volume 2017**.

52) J. Zhu, D. Li, H. Chen, X. Yang, L. Lu, X. Wang, 2004, "Highly dispersed CuO nanoparticles prepared by a novel quick-precipitation method", *Materials Letters*, **58**, 3324 – 3327.

Acknowledgments

I acknowledge my deep sense of gratitude to my research advisors Dr. Raminder Kaur & Dr. Poonam Singh, Assistant Professor, Department of Applied Chemistry, Delhi Technological University, Delhi for their guidance, motivation and remarkable patience throughout this research work. Also, I am grateful to the research scholars, Ms. Bhamini Pandey & Ms. Jigyasa Pathak, and the technical staffs of Department of Applied Chemistry, Delhi Technological University, Delhi for their constant support and assistance during this research work.

Effect of Silica Fume & Steel Slag on Nano-silica based High-Performance Concrete

C M Kansal[1], S Singla[1*] and R Garg[2]

[1]Department of Civil Engineering, RIMT University, Mandi Gobindgarh, India
[2]Department of Civil Engineering, Galgotias College of Engineering and Technology, Greater Noida, India
*Corresponding author: drsandeepsinglaz@gmail.com

Abstract. This study presents strength behaviour of concrete based composites including three different ternary binders namely Nano-Silica (NS), Silica Fume (SF) & Steel Slag (SS). NS as fragmental substitution of cement, SF as the fragmental substitution for the fine aggregates and SS as fragmental substitution of coarse aggregates has been used. The study has been conducted at an optimized content of NS (2%) and by varying the content of SF (10-15%) and SS (10-30%). Cement concrete having M60 grade strength at fixed W/C (0.35) has been explored for the strength behaviour. The studies have revealed that the optimized content of NS, SF and SS can be used as a partial substitute of cement, fine aggregates and coarse aggregates respectively to obtain increased strength. Thus, NS, SF and SS can be used to obtain high performance and eco-friendly concrete structures.

Keywords: *Steel Slag, Silica Fume, Nano silica, Concrete, Compressive Strength.*

1. Introduction

India is a country, which is interested in leading the world in globalization, and therefore the country should continuously develop itself especially by enhancing industrialization and innovative construction technology. The growing concern over detrimental environmental impact of construction industry in terms of responsive carbon dioxide emissions as well as durability concerns of structures, researchers are looking after innovative solutions [1]. Over time, high-performance concrete (HPC) has been developed and undergone many ameliorations for improvements. The concrete with strength more than M55 grade is commonly known as high strength concrete [2]. The use of HPC is increasing day by day because of its excellent structure performance and environmental friendliness with lesser impact on energy utilization [3]. HPC can withstand the impact of high temperature and maintain its toughness and strength due to presence of strength modifiers [4]. Supplementary cementitious materials (SCMs), namely silica fume (SF) [5], nano silica (NS) [6], ground granulated blast furnace slag (GGBS) [7], steel slag (SS) [8] and fly ash (FA) [9] are the essential substances which are used for fabrication of HPC. When such ingredients are used as additives for HPC, the strength and bending characteristics of concrete are enhanced [10]. Especially, the SCMs such as pozzolans including FA, SF and NS are highly desirable as these materials enhance the strength of the concrete by propagating the formation of additional calcium silicate hydrate (CSH) gel, the prime factor for strength enhancement [11]. With reference to worldwide concern over sustainability, the due utilization of industrial waste is highly desirable [12]. Researchers are working continuously on the performance of high strength concrete using mineral admixtures/Industrial waste in place of regular main/natural ingredients [13]. The best possible route could be the utilization of such waste and conversion into a usable product that can be reused as raw materials or as strength modifying materials [12]. Ecologically sustainable HPC has been obtained by using recycled concrete aggregates in presence of fly ash and has been observed to exhibit high strength and durability in reinforced structures [14]. It has been reported that when cement is partially replaced with strength enhancers, the improved strength properties have been observed [15].

Steel slag is a by-product of steel industry that is obtained as a molten liquid when iron is converted into steel or by melting steel in an electric arc furnace [16]. The molten liquid, which is a complex silicate and oxide solution solidifies and turns into a steel slag after the cooling process. Steel slag is a non-metallic product made up of calcium and ferrite that is mixed with iron, aluminium, magnesium, calcium, magnesium oxide and is merged among steel in an electric arc or open-hearth furnace with the help of basic oxygen. Use of steel slag has been reported to produce eco-friendly HPC as demonstrated by life cycle analysis and toxicity investigations [8]. Steel slag has been used as a substituent of coarse aggregates at a lower water/cement ratio ranging in between 1.76-2.21 and coarse/fine aggregate ratio as 1.98. An early and significant strength enhancement was recorded for the specimens [17]. Silica fume is a by-product of silicon-based alloy industry and is primarily non-crystalline (amorphous) silicon dioxide (usually over 85%) with particle size < 1 μm in diameter [18]. Its particles are 100 times finer than the average cement particles. Silica fume is used in between 5-15 % by the weight of the cement or fine aggregates. It is used in places where concrete with high impermeability, durability and high strength is required [19]. Incorporation of SF in concrete at a varying range of 5-25% has been observed to significantly enhance the strength aspects of concrete and best results were obtained at 10% content of SF [5]. Nano-silica is mainly used as a pozzolanic and antibleeding agent in HPC and SCC concrete. Its incorporation makes concrete more cohesive and the segregation tendency is reduced. According to literature, the addition of colloidal nano-silica (0.5-2% by weight of cement) in concrete and mortar will ultimately enhance the strength [20]. Similarly, the use of colloidal nano-silica at a range of 0.5-2.0% in HPC concrete was found to impart accelerated hydration, and improved chloride ion resistance at low water/binder ratio [21]. A remarkable decreased chloride permeability was attributed to the increased compactness of the interfacial transition zone [22].

The introduction of binary pozzolanic systems at optimized content has been found to impart better results ascribed to the enhanced pozzolanic reaction [23]. Hence, a combination of three SCMs, namely NS, SF and SS to partially substitute the binders in HPC can produce better results. With this consideration, the main objective of this study is partial replacement of cement as well as fine and coarse aggregates by NS, SF and SS respectively to develop high strength and ecologically sustainable concrete. The compressive strength, split tensile strength, and flexural strength has been investigated. The research provides a systematic scientific analysis utilizing admixtures and analyzing their effects.

2. Materials and Methods

This examination was carried out using Ordinary Portland Cement (OPC) 43 grade (Ultra Tech Cement) from a solitary batch. NS, SF and SS as well as the fine and coarse aggregates were procured from the

IOP Conf. Series: Materials Science and Engineering **961** (2020) 012012 doi:10.1088/1757-899X/961/1/012012

local suppliers. The physical properties and chemical composition of the components has been listed in Table 1 and Table 2 respectively. Steel Slag and Silica Fume are similar in size to coarse and fine aggregates respectively making the concrete mixture to be operational or workable. However, the use of Steel Slag and Silica Fume may negatively impact the workability of the concrete combination, and consequently, water-reducing admixtures were mixed. The mix composition has been listed in Table 3 for the seven mix specimens at water/binder ratio of 0.35 and in presence of superplasticizer. The specimens were prepared as control specimen with only 2% NS (C0), specimen with 2% NS and 10% SF with incorporation of 10% SS (M1), 20% SS (M2), 30% SS (M3) and specimen with 2% NS and 15% SF with incorporation of 10% SS (M4), 20% SS (M5), 30% SS (M6). The strength analysis including compressive strength, flexural strength, and split tensile strength was carried out at 7, 28 and 56 days [15].

Table 1. Physical Properties of Components

Aggregate Type	Bulk Density (kg/m^3)	Fineness modulus	Specific gravity
NS	<100	-	1.25
SS	1620	5.43	2.90
SF	1465	2.36	2.22
Fine aggregates	1615	2.83	2.55
Coarse aggregates	1527	5.52	2.74

Table 2. Chemical Composition of Binder Materials

Oxides % by mass	OPC % by mass	NS % by mass	SF % by mass	SS % by mass
SiO_2	19.46	99.9	98	26.4
Al_2O_3	4.22	-	0.03	4.84
Fe_2O_3	3.56	-	0.02	43.4
CaO	65.92	-	0.3	16.9
MgO	1.08	-	0.2	1.86
K_2O	0.67	-	0.3	0.11
Na_2O	0.21	-	0.3	0.15
TiO_2	0.24	-	-	0.22
SO_3	2.6	-	-	2.58

Table 3. Mix proportion of Binders per m^3

Designation	Cement (g)	NS (g)	SF (g)	SS (g)
C0	1000	20	-	-
M1	980	20	100	100
M2	980	20	100	200
M3	980	20	100	300
M4	980	20	150	100
M5	980	20	150	200
M6	980	20	150	300

3. Results and Discussions

3.1. Compressive Strength

Concrete compressive strengths of specimens C0, M1, M2, M3, M4, M5 and M6 at 7, 28, and 56 days has been illustrated in fig.1. All the ternary specimens were found to exhibit higher strength as compared to the control specimen. The increased compressive strength on substituting the fine aggregates with SF and coarse aggregates with SS is due to the densification of the structure of the matrix because of the accelerated pozzolanic action [4]. The fine silica particles of SF occupy the pores in the concrete matrix and also react with calcium hydroxide to generate additional CSH gel. Likewise, the calcium oxide and silica particles in steel slag are strength enhancing in nature and improve the strength. Further, there is a decrease in fine aggregates and coarse aggregates content, in turn increasing the content of SCMs. This leads to better binding between the binder materials with reduced porosity of the matrix. The synergistic effect of SF in presence of SS provide better results as compared to C0 [16].

Concrete Mixture M2 with 2.0% NS, 10% SF and 20% SS was found to exhibit the maximum compressive strength at all the studied ages. Increasing the amounts of SS from 10% to 20% at constant SF content was found to increase the compressive strength. However, further increase of SS content to 30% at constant SF content was found to result in a comparative decrease in the strength. The percent increase in compressive strength as compared to C0 was observed as 3.0%, 10% and 6.0% at 28 days for M1, M2 and M3 specimens. On the other hand, the percent increase in compressive strength for M4, M5 and M6 specimens at 28 days was observed as 2.0%, 6.0% and 3.0% respectively as compared to C0. The slight reduction in strength at increasing content of SS and SF is possibly due to increase in friction among the particles.

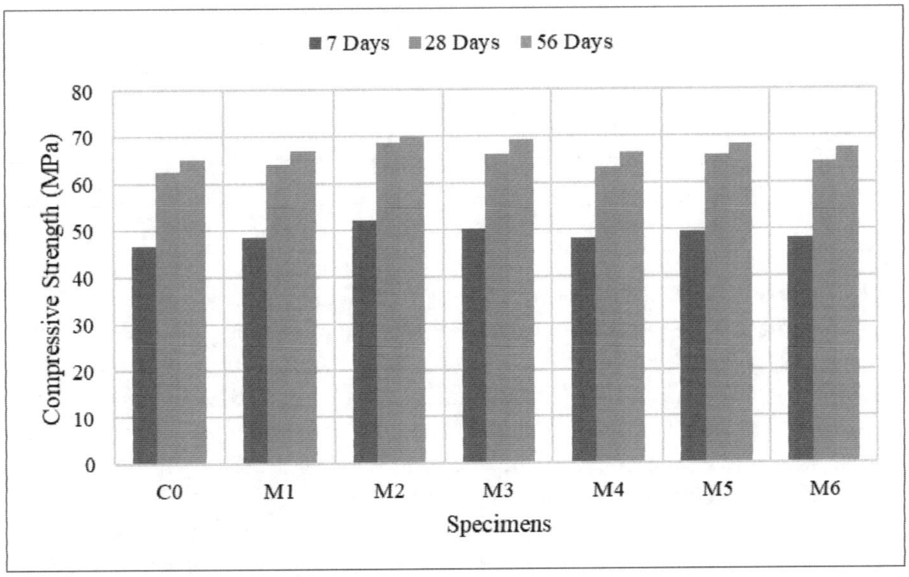

Figure 1. Compressive Strength of Specimens

3.2. Split Tensile Strength

Splitting tensile strength of the concrete specimens was measured at 7, 28 and 56 days and the results have been illustrated in fig.2. Splitting tensile strength of C0 was observed as 4.10 MPa on 7 days, 5.36 MPa on 28 days and 6.13 MPa on 56 days. With the presence of Silica Fume & Steel Slag at all ages, the splitting tensile strength was found to increase as compared to C0. The behavior was similar to compressive strength results with presence of SF and SS. The percent increase in split tensile strength as compared to C0 was observed as 6.0%, 13% and 9.0% at 28 days for M1, M2 and M3 specimens. On the other hand, the percent increase in split tensile strength for M4, M5 and M6 specimens at 28 days was observed as 5.0%, 9.0% and 6.0% respectively as compared to C0. The slight decrease in strength was attributed to the loss of binding at higher concentration of SF and SS due to increased repulsive forces between the finer particles [10]. However, the results were significantly improved as compared to C0 indicating the suitability of SF as a partial substitute of fine aggregates and SS as a partial substitute of coarse aggregates.

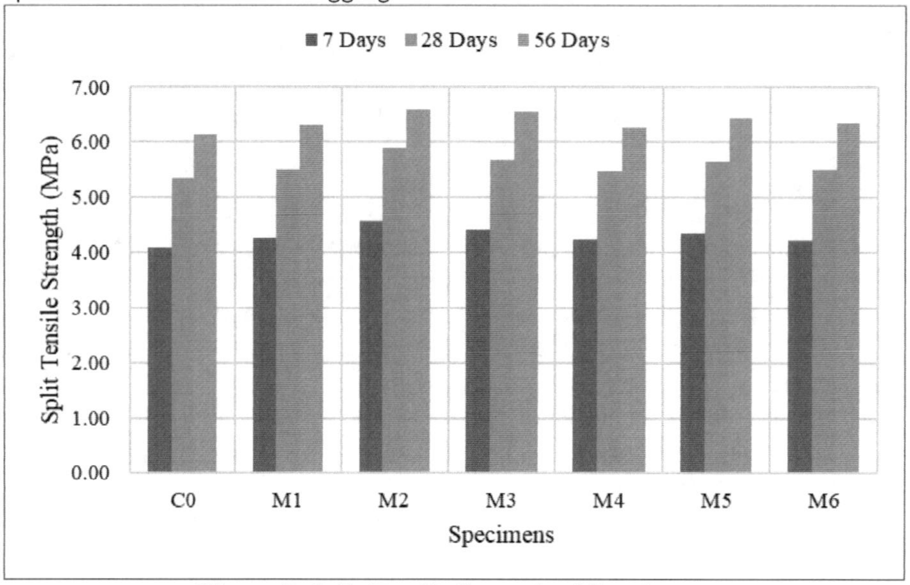

Figure 2. Split Tensile Strength of Specimens

3.3. Flexural Strength

Flexural strength of the concrete specimens was measured at 7, 28 and 56 days and the results have been illustrated in fig.3. Flexural strength of C0 was observed as 7.26 MPa on 7 days, 7.91 MPa on 28 days and 8.23 MPa on 56 days. The flexural strength of all specimens were found to increase with presence of SF and SS as compared to C0. The percent increase in flexural strength as compared to C0 was observed as 5.0%, 8.0% and 7.0% at 28 days for M1, M2 and M3 specimens. On the other hand, the percent increase in flexural strength for M4, M5 and M6 specimens at 28 days was observed as 6.0%, 7.0% and 5.0% respectively as compared to C0. The substitution of fine aggregates with SF upto 10% in presence of 2.0% NS and 20% SS as a partial substitute of coarse aggregates was found to exhibit the maximum flexural strength at all the studied ages. The increase in SF and SS content beyond this content was found to cause slight decrease in strength. The amount of air voids in the matrix is one of the crucial elements of the concrete mixture [5]. Air voids help water to drain into pores, reducing high internal stress and structural micro cracking. As the air voids are occupied at higher content of steel slag and silica fume, the decreased availability of water may decrease the hydration reaction causing negative impact on the concrete strength [22].

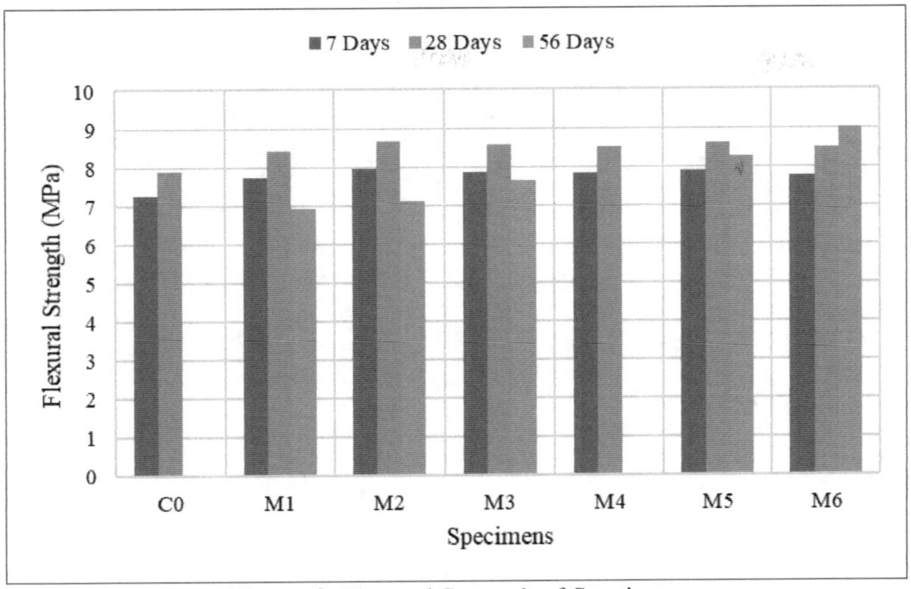

Figure 3. Flexural Strength of Specimens

4. Conclusion

The goal of this research was to study the strength behaviour of high-performance concrete, by replacing the coarse aggregates and fine aggregates by two industrial waste products namely silica fume (SF) and steel slag (SS). Enhanced compressive strength, flexural strength and split tensile strength was observed for these specimens indicating their effectiveness as a partial substitute in the concrete mixes. Hence, the use of SS and SF as a partial substituent of natural aggregates not only reduces the environmental impact of these wastes, but also reduces the global carbon dioxide emission. The findings showed that the strength characteristics of high-performance concrete by using these industrial wastes exhibit significant results up to 10% SF and 20% SS, as compared to conventional concrete with no negative impact on workability in presence of superplasticizer. Thus, an optimized use of SS and SF in presence of NS can significantly improve the strength aspects of concrete structures as well as help in sustainable construction practices.

References

[1] Garg R and Garg R 2020 Effect of zinc oxide nanoparticles on mechanical properties of silica fume-based cement composites *Mater. Today Proc.*

[2] Gülşan M E, Alzeebaree R, Rasheed A A, Niş A and Kurtoğlu A E 2019 Development of fly ash/slag based self-compacting geopolymer concrete using nano-silica and steel fiber *Constr. Build. Mater.* **211** 271–83

[3] de Matos P R, Oliveira J C P, Medina T M, Magalhães D C, Gleize P J P, Schankoski R A and Pilar R 2020 Use of air-cooled blast furnace slag as supplementary cementitious material for self-compacting concrete production *Constr. Build. Mater.* **262**

[4] Chen H J, Yu Y L and Tang C W 2020 Mechanical properties of ultra-high performance concrete before and after exposure to high temperatures *Materials (Basel).* **13** 1–17

[5] Smarzewski P 2019 Influence of silica fume on mechanical and fracture properties of high performance concrete *Procedia Struct. Integr.* **17** 5–12

[6] Garg R, Bansal M and Aggarwal Y 2016 Strength, rapid chloride penetration and microstructure study of cement mortar incorporating micro and nano silica *Int. J. Electrochem. Sci.* **11** 3697–713

IOP Conf. Series: Materials Science and Engineering **961** (2020) 012012 doi:10.1088/1757-899X/961/1/012012

[7] Lei Voo Y and Foster S J 2010 Characteristics of ultra-high performance “ductile” concrete and its impact on sustainable construction *IES J. Part A Civ. Struct. Eng.* **3** 168–87
[8] Zhang X, Zhao S, Liu Z and Wang F 2019 Utilization of steel slag in ultra-high performance concrete with enhanced eco-friendliness *Constr. Build. Mater.* **214** 28–36
[9] Yildiz K A and Uğr L O U 2009 The effect of mineral admixture type on the modulus of elasticity of high strength concrete *Sci. Res. Essays* **4** 791–8
[10] Falmata A M, Sulaiman A, Mohamed R N and Shettima A U 2020 Mechanical properties of self-compacting high-performance concrete with fly ash and silica fume *SN Appl. Sci.* **2** 1–11
[11] Garg R, Garg R, Bansal M and Aggarwal Y 2020 Experimental study on strength and microstructure of mortar in presence of micro and nano-silica *Mater. Today Proc.*
[12] Shehab El-Din H K, Eisa A S, Abdel Aziz B H and Ibrahim A 2017 Mechanical performance of high strength concrete made from high volume of Metakaolin and hybrid fibers *Constr. Build. Mater.* **140** 203–9
[13] Zain M F M, Safiuddin M and Mahmud H 2000 Development of high performance concrete using silica fume at relatively high water-binder ratios *Cem. Concr. Res.* **30** 1501–5
[14] Kubissa W, Simon T, Jaskulski R, Reiterman P and Supera M 2017 Ecological High Performance Concrete *Procedia Eng.* **172** 595–603
[15] Garg R and Garg R 2020 Performance evaluation of polypropylene fiber waste reinforced concrete in presence of silica fume *Mater. Today Proc.* 6–13
[16] Furlani E, Tonello G and Maschio S 2010 Recycling of steel slag and glass cullet from energy saving lamps by fast firing production of ceramics *Waste Manag.* **30** 1714–9
[17] Nguyen T T H, Phan D H, Mai H H and Nguyen D L 2020 Investigation on compressive characteristics of steel-slag concrete *Materials (Basel).* **13**
[18] Vishwakarma V and Ramachandran D 2018 Green Concrete mix using solid waste and nanoparticles as alternatives – A review *Constr. Build. Mater.* **162** 96–103
[19] Raheem A H A, Mahdy M and Mashaly A A 2019 Mechanical and fracture mechanics properties of ultra-high-performance concrete *Constr. Build. Mater.* **213** 561–6
[20] Garg R, Bansal M and Aggarwal Y 2016 Split Tensile Strength of Cement Mortar Incorporating Micro and Nano Silica at Early Ages *IJERT* **5** 16–9
[21] Du H and Pang S D 2019 High performance cement composites with colloidal nano-silica *Constr. Build. Mater.* **224** 317–25
[22] AlKhatib A, Maslehuddin M and Al-Dulaijan S U 2020 Development of high performance concrete using industrial waste materials and nano-silica *J. Mater. Res. Technol.* **9** 6696–711
[23] Liu H, Zhang Y, Tong R, Zhu Z and Lv Y 2020 Effect of Nanosilica on Impermeability of Cement-Fly Ash System *Adv. Civ. Eng.* **2020**

IOP Conf. Series: Materials Science and Engineering **961** (2020) 012013 doi:10.1088/1757-899X/961/1/012013

Mechanical Properties of Chemically Treated Banana and Ramie Fibre Reinforced Polypropylene Composites

G.Sai Krishnan[1], G.Shanmugasundar[2], M.Vanitha[3], N.Sivashanmugam[1]

[1]Department of Mechanical Engineering, Rajalakshmi Institute of Technology
[2]Department of Mechanical Engineering, Sri Sairam Institute of Technology
[3]Department of Chemistry, Sri Sairam Engineering College
*Corresponding author: g.saikrishnan@gmail.com

Abstract. Present research focuses on finding better fibre reinforcement cheap and eco-friendly polypropylene composite Mechanical Properties increased with increase in fibre loading and sodium carbonate Treatment. Banana and Ramie Fibres were chosen as reinforcement .The compression moulding technique was used to prepare the composite material. Fibre loading was varied at 0, 5 and 10 % volume. Both raw Banana and Ramie Fibre was chemically treated with 5 wt% sodium carbonate and composites were also prepared with 0, 5 and 10% volume treated ramie and banana fibre. Analysis of prepared composites was subsequently conducted using Mechanical Tensile, flexural, hardness tests and. 10% fibre reinforced composite provided the best set of mechanical properties. It can be concluded after chemical treatment the mechanical treatment got enhanced.

Keywords: Banana Fibers, Ramie Fibers, Chemical Treatment, Polypropylene Composites.

1. Introduction

The replacement of polymeric fibers for existing synthetic fibers is in high demand owing to its applications. The main reason for this is due to easy availability, geographical location and easy accessibility. The processing of these fibers is also easy. Many of the modern technologies especially aerospace and transportation require the materials having the properties combination of metal, ceramics and polymer. The natural fibers are used as an key ingredient in various applications such as automobile, household and aero space[1-6]. Polymer matrix composites reinforced by various synthetic fibres fulfil the most of the demands. However their high processing cost, less availability, non-biodegradability and health hazardous are the main matter of concern. One of the important characteristics of natural fibers are inexpensive and will control the environmental side affects [7]. Out od the fibers such a glass,synthetic and plant based fibers.The utilization of plant based fibers increased day by day due to its ease extraction and availability.Chemical analysis is one of he important paramaeters of analysing the quality of fibers.The cellulose and amrphous contents are vital in examing he mechanical and morphological properties of the fibers.The various amorphous contents such as hemicellulose,lignin are removed to improve the mechanical performance.The strength is also

dependent on the crystalline (cellulose) and amorphic (hemicellulose, lignin, pectin) contents which are directly dependent on the environmental conditions and location in which the plant is present.
The chemical treatment becomes effective only when the optimal chemical concentration and soaking time are followed, else there will be drastic removal of chemical constituents from the composition of fibre thereby leading to poor results. There are many chemical treatments that are widely used namely alkalization, benzoyation, peroxide, stearic acid, HCl, stearic acid, silane treatment, etc for surface modification of the fibres..Banana and Ramie fibre reinforced polypropylene composites were prepared using hand layup technique After obtaining they were treated by sodium carbonate (Na_2CO_3).In order to increase the adhesion between polypropylene and fibre. Mechanical testing and thermal characterization of prepared composites were conducted. Finally the effect of fibre type, fibre loading and chemical treatment on mechanical and thermal properties was evaluated.

2. Preparation of Composites

The banana and ramie fibers were prepared by hand layup techniques. The fibers were retted and developed as a composite. Initially the fibers were collected from local market in Chennai and soaked in water for 24 hours. Later the chemical treatment was performed. Sodium carbonate treatment was carried out for the retted fibers.The same step by step procedure was carried out which was done in our previous literature[8-10].

3. Results and Discussion

3.1 Banana Fibre and Ramie Fibre

The Surface modification of Banana Fibre and Ramie Fibre was done by using Na_2CO_3.It seems that both banana and ramie fibres after modifications had some smooth surfaces with less impurities and wax or fatty layer. The Treatment of fibres decreased the spiral and increased the orientation of fibres. Because of roughening, bonding between fibre and matrix increased after chemical treatment. The mechanical properties of the fibres were increased and the details were mentioned below. The retted fibers images were shown below.

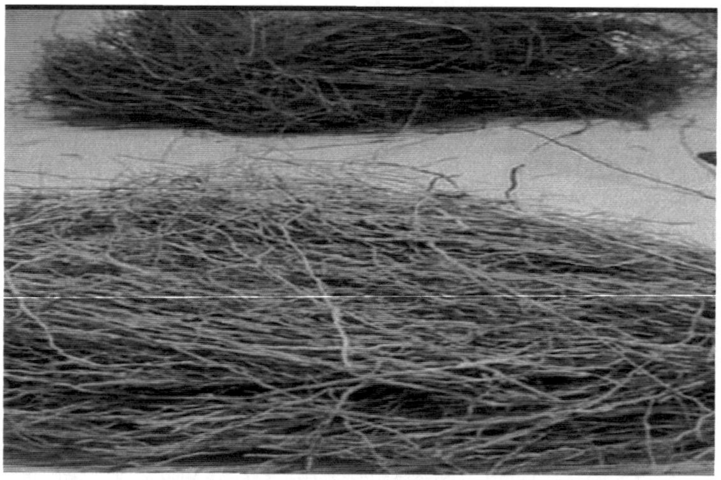

Figure1:Images of Retted Fibers

3.2 Tensile Properties

Tensile strength of polypropylene and various prepared composites is shown in Figure . Tensile Strength of prepared composites was higher than that of pure polypropylene. Previous research Also confirms the same trend [11]. Tensile strength also increased with increase in fibre loading. As we know that both the banana and hemp fibres are having good tensile strength when it is mixed with polypropylene. Thehigher fibre ratio banana and hemp fibres increased the fibre tensile strength. On comparing these two fibres banana had good results than the hemp because of its better reinforcements when compared to hemp fibre. The Tensile Strength of the banana fibres were increased from 23.5% to 34.56 % for 5 and 10 weight percentages. Similarly on the other hand tensile strength of the ramie fibres was in the range of 31.23% which is lower than that of the banana fibres. This is mainly due to the presence of uniform cross section of banana fibre to the aspect ratio. Banana fibre has more uniform cross section and greater aspect ratio as compared to Ramie fibres. Treatment of banana fibres with Na_2CO_3 gave the better results compared to ramie. This is mainly due to the distribute load of the matrix in a better way. It also restricts the growth and formation of crack which will restricts the failure better. The outcome of this results in a banana fibre with increased tensile strength. The same trend was also found in the work carried out before[12-14].

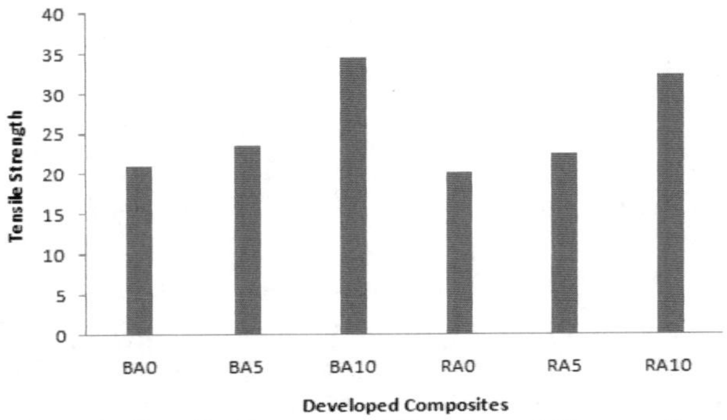

Figure2: Tensile Strength of Developed Composites

3.3 Flexural Properties

The Flexural Strength of the developed composites is analysed and the details are given below. Flexural strength varies with the similar fashion as tensile strength. Flexural strength ofpolypropylene and various prepared composites is shown in Figure 2. Flexural strength increasedwith an increase in fibre loading, which is in agreement with the findings of other researchers [15].Incorporation of 10% banana fibre in polypropylene increased flexural strength by 7.23 percentages and banana fibres with 10 and 15 weight percentages increased the flexural strength by 25.3% and 32.7%. The favourable entanglement of the polymer chain with the filler may be the reason of improvement . 5% Na_2Co_3 treatment increased flexural strength of both jute and banana fibrereinforced polypropylene composites as compared to raw jute and banana fibre reinforcedpolypropylene composites. Previous research confirmed this phenomenon due to surfacemodification of the fibre [18-20].There is not much deviation between the banana and ramie fibres this may be attributed to better adhesion between the fibres and the matrix[16-18].

IOP Conf. Series: Materials Science and Engineering **961** (2020) 012013 doi:10.1088/1757-899X/961/1/012013

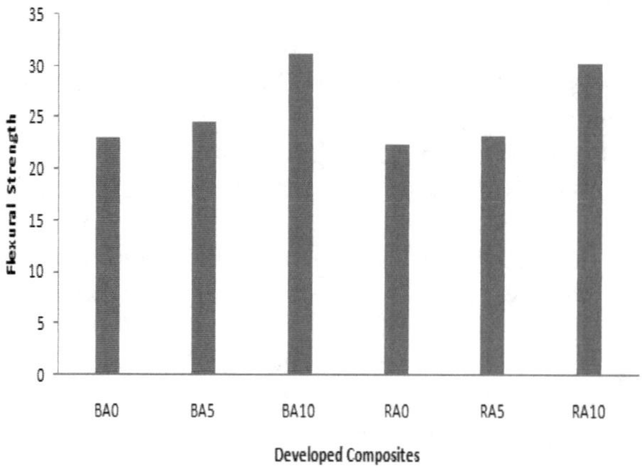

Figure3: Flexural Strength of Developed Composites

3.4 Hardness

Hardness of polypropylene and various prepared composites is shown in Figure 4. Shorehardness of pure polypropylene was found to be 94.6. Due to fibre addition and fibre treatmentshore hardness increased significantly [19]. Hardness of 5% Na_2CO_3 treated banana fibre was 95.2 whereas 10% had some increase in the value of 97.45.Increasing the fibre loading increased the hardness of the composites. On the other hand,5% of ramie had hardness of 93.25 whereas hardness 10% of ramie fibres was 96.8. This scenario could be linked to the better adhesion of the matrix to the fibre. The addition of natural fibre to a polymer matrix has-been seen to reduce flexibility by retarding the molecular movement of the polypropylenematrix.Fibre addition also decreases the flexibility of the matrix and increases hardness [20-21].

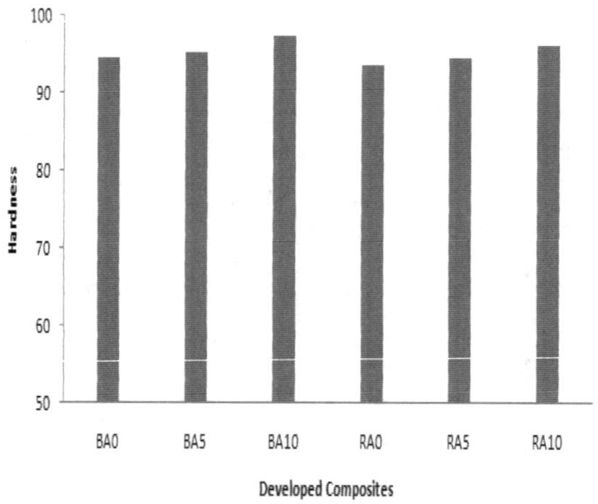

Figure 4: Hardness of Developed Composites

4.Conclusion

In present research, banana fibre and Ramie Fibres were prepared composites were prepared with 0, 5 and 10 volume% fibre lading. Both Ramie and banana fibre was treated with 5% Na_2CO_3 solution.The various Properties such as tensile, Flexural and hardness got enhanced by using Chemical treatments 10% banana fibre reinforced composite provided the best set of mechanical properties. Depending on these experimental results, one can easily produce low cost environment friendly high strength polymer matrix composites for automobile industry as well as domestic purpose. In future it can be utilised for automobile applications.

5.References

[1] Jafrey Daniel D, Sai Krishnan G and Velmurugan PInvestigation on the characteristics of bamboo/jute reinforced hybrid epoxy polymer composites Mater. Res. Express 6 105346
[2] Sai Krishnan G, Jayakumari L S, Babu L G and Suresh G (2019) Investigation on the physical, mechanical and tribological properties of areca sheath fibers for brake pad applications Mater. Res. Express 6 085109
[3] C.Sivakandhan, Ganesh Babu Loganathan (2020) Material characterization and unconventional machining on synthesized Niobium metal matrix Mater. Res. Express 7 015018

[4] James J D, Manoharan S, Saikrishnan G and Arjun S (2019) Influence of bagasse/sisal fibre stacking sequence on the mechanical characteristics of hybrid-epoxy compositesJ. Nat. Fibers 2 1–11
[5] Saikrishnan G, Jayakumari L S, Vijay R and Lenin Singaravelu D (2019) Influence of iron–aluminum alloy on the tribological performance of non-asbestos brake friction materials–a solution for copper replacement Industrial Lubrication and Tribology 00 1–13
[6] Manoharan S, Shihab A I, Alemdar A S A, Ganesh Babu L, Vijay R and Lenin Singaravelu D 2019 Influence of recycled basalt-aramid fibres integration on the mechanical and thermal properties of brake friction composites Material Research Express 6 115310
[7] G Sai Krishnan, S Kumar (2020) Role of metal composite alloys in non-asbestos brake friction materials-A solution for copper replacement Materials Today: Proceedings 2020/3/31
[8] Manoharan S, Sai Krishnan G, Babu L G, Vijay R and Singaravelu D L 2019 Synergistic effect of red mud-iron sulfide particles on faderecovery characteristics of non-asbestos organic brake friction composites Mater. Res. Express 6 105311
[9] Jawaid, M., and Khalil, H. A. (2011) Effect of layering pattern on the dynamic mechanical properties and thermal degradation of oil palm-jute fibers reinforced epoxy hybrid composite. Bio Resources, 6(3): 2309-2322.
[10] G. Sai Krishnan(2019) Study on mechanical & thermal properties of PCL blended graphene bio composites Polimeros1-9
[11].GS Krishnan(2020) Investigation of Caryota urens fibers on physical, chemical, mechanical and tribological properties for brake pad applications Materials Research Express 015310
[12] G Sai Krishnan(2020) Investigation on the physical and mechanical characteristics of demostachya bipinnata reinforced with polyester composites Materials Today: Proceedings 2020/4/
[13] G.Sai Krishnan Babu L G and Pradhan R (2020) Study on tribological properties of palm kernel fiber for brake pad applications Materials Research Express Mater. Res. Express 7 015102
[14] Nongman. A. Fa, A. Baharina, and A.AbuBakarb ,—The Effect of Banana Leaves Lamination on the Mechanical Properties of Particle Board Panel‖ 5th International Conference on Recent Advances in Materials, Minerals and Environment (RAMM) & 2nd International Postgraduate Conference on Materials, Mineral and Polymer (MAMIP), 4-6 August 2015
[15] Ramnath, B. V., Kokan, S. J., Raja, R. N., Sathyanarayanan, R., Elanchezhian, C., Prasad, A. R., and Manickavasagam, V. M. (2013) Evaluation of mechanical properties of abaca–jute–glass fibre reinforced epoxy composite. Materials & Design, 51: 357-366.

[16] Vinod A, Vijay R and Singaravelu DL. (2018) Thermomechanical characterization of Calotropis gigantea stem powder-filled jute fiber-reinforced epoxy composites. J Nat Fibers Journal of Natural Fibers; 15: 648–657.
[17] Vijay R and Singaravelu DL(2016). Experimental investigation on the mechanical properties of Cyperuspangorei fibers and jute fiber-based natural fiber composites. International Journal of PolymerAnalysisi and Characterization 2016; 21: 617–627.
[18] Tej Singh Sachin Tejyan Amar Patnaik(2019) Fabrication of waste bagasse fiber-reinforced epoxy composites: Study of physical, mechanical, and erosion propertiespolymer composites—2019
[19] Maha M. Ibrahima, Alain Dufresneb, Waleed K. ElZawawya,*, Foster A. Agblevorc —Banana fibres and micro fibrils as lignocellulosic reinforcements in polymer composites‖ Article in Carbohydrate Polymers July 201
[20] Amira,*, Kamal AriffZainalAbidinb,FaizzatyBintiMdShiric —Effects of Fibre Configuration on Mechanical Properties of Banana Fibre/PP/MAPP Natural Fibre Reinforced Polymer Composite‖ Advances in Material & Processing Technologies Conference
[21] Jothibasu, S., Mohanamurugan, S., & Vinod, A.(2017) Influence of chemical treatments on the mechanical characteristics of areca sheath-flax fibres based epoxy composites. Rasayan Journal of Chemistry, 11(3), 1255-1262

ANALYSIS AND OPTIMIZATION OF MACHINING PARAMETER DURING TURNING OF A356/10%SiCp MMC USING RESPONSE SURFACE METHODOLOGY APPROACH

M. Seeman[a*], P. Sivaraj[b], R. Seetharaman[c] and I. Ashok Kumar[d]

[a, b]Department of Manufacturing Engineering, Annamalai University, Annamalainagar-608 002, Tamilnadu, India.

[c]Department of Mechanical Engineering, University College of Engineering, Anna University, Dindigul – 624 622, Tamilnadu, India.

[d] Department of Mechanical Engineering Government Engineering College, Bargur, Krishnagiri-635104, Tamilnadu, India.

Corresponding author Tel.: 09443673644, +91 4144239734; Fax: +91 4144239734.

Email address: mfgseeman@gmail.com, seeman_au575@rediffmail.com (M.Seeman)

ABSTRACT

In this study, an effort has been made more complex multi performance optimization using Response Surface Methodology (RSM) based desirability approach while machining of Al/10%SiC Particulate Aluminum Metal Matrix Composite (PAMMC) produced through stir cast method. The effect of main cutting parameters like Cutting speed (V), Feed rate (F), Depth of cut (D) and Machining time (T) on the output responses of Tool flank wear (VB_{max}) and Surface roughness (Ra) were investigated. The most significant machining parameters were ascertained using analysis of variance (ANOVA). The contour plots are drawn for study effect cutting parameters and their interactions. From the study found that developed numberical model for tool flank wear and surface roughness is found to be adequate and used for predict the parameters within this experimental range.

Keywords: Particulate Aluminum Metal Matrix Composite (PAMMC); Response surface methodology (RSM); Analysis of Variance (ANOVA); Tool flank wear (VB_{max}); Surface roughness (Ra); Contour plots.

IOP Conf. Series: Materials Science and Engineering **961** (2020) 012014 doi:10.1088/1757-899X/961/1/012014

1. INTRODUCTION

Metal matrix composites (MMC's) are one among the commonly well-known composites material due to their superior mechanical and chemical properties such as high hardness, strength, toughness, wear resistance and corrosion resistance. Silicon carbide particulate (SiCp) reinforced aluminium metal matrix composites are most widely used MMC and commercially available ones due to their cost-effective production processes [1]. Mostly the products made out of composite materials are near-net shaped processes, but the need for machining cannot be completely eliminated, since, in modern engineering, high demands are being placed on components made of composites in relation to their dimensional precision as well as their surface quality [2]. Therefore, there is a great need to understand the questions associated with machining of this type of composite material.

Progressive technologies need materials with different exceptional combinations of properties that cannot be met by the conventional metal, alloys, ceramics, and polymeric materials. This is especially true in the case of materials that are needed for aerospace and space industries, where the product performance requirement is quite complex. The demands made on materials for better overall performance are so great and diverse that no single material can satisfy them. Hence, new materials have to be synthesized, which ultimately result development of advanced composite materials [3].

The industrial application Al/SiC particulate MMC still inadequate this mainly due difficulty in machining presence of abrasive nature of hard SiC particles leads to excess wear on tool. From the published research found that only PCD tool gives the use full tool life while machining Al/SiCp MMC due their high hardness and chemical propensity. But considering cost of machining carbide and ceramic cutting tools are not completely eliminated. The carbide cutting tools have useful tool life in lower cutting speeds less than 75 m/min. So many researcher recommend cemented carbide cutting tools for rough cutting operation and PCD tool for finishing in order to minimizing machining cost [4].

However, a limited quantity of research are stated on the machining A356 Al/SiCp MMC with uncoated carbide cutting tools respect to output responses of tool flank wear and surface roughness using response surface methodology with different cutting conditions [5, 6, 7 & 8]. Moreover, the significant number of studies has been reported on machining different other alloys of aluminium as matrix and SiC has been used as reinforcement using carbide, CBN and PCD cutting tools [9, 10, 11, 12, 13,14, 15, 16 &17]. Hence the main objective of the present work is to optimize tool wear and surface roughness while machining LM25 Al/SiCp metal matrix composite using uncoated cemented carbide cutting tools (K10) with Response surface methodology (RSM) based desirability approach.

2. EXPERIMENTAL STUDY

Turning experiments were performed on a center lathe (PSG 141) which is having a spindle speed range of 50-1750 rpm with feed range 0.05-3.5 mm/rev with adequate spindle control. A356 equivalent of British grade LM25 Al alloy chemical composition as shown in Table 1 is reinforced with 10% SiC particles with an average grain size of 25μm was produced by stir casting method is used for conduct the turning trials. The dimensions A356/SiC/10p MMC rod used for experiment was 275 mm length and 100 mm diameter. The uncoated fine grain cemented carbide cutting tool ISO grade of K10 grade WIDIA make used for experiment. The turning trials were conducted as per DOE (Design of Experiment) matrix at random manner to eliminate logical mistakes.

To measure the tool flank wear the maximum width of flank (VBmax) land was measured in this study which is widely used in manufacturing applications in the industry. The maximum width of flank (VBmax) and Build up Edge (BUE) formation quantified using CLEMAX optical microscope with following specification: Working distance: 1 to 5 mm, Magnification: 50 to 1000X, Illumination: 12V, 100W. The Scanning Electron Microscope (SEM) was used for analyzing tool flank wear mechanism at various machining parameters.

Table 1 Elements of A356 Al alloy

Elements	Si	Fe	Mn	Mg	Ti	Ni	Cu	Al
A356 Al alloy	7	0.5	0.3	0.33	0.2	0.1	0.1	Bal

3. DESIGN OF EXPERIMENTS BASED ON RESPONSE SURFACE METHODOLOGY

In this study, main cutting conditions like Cutting speed (V), eed rate (F), Depth of cut (D) and Machining time (T) was considered as important design factors in the turning process. To study the performance of the tool flank wear (VB_{max}) and surface roughness (Ra) were considered as out responses.

The experiments were designed using full factorial face centered CCD (Central Composite Design) mixing of all factors of cutting parameters in two levels high (+1) and low (-1) consist of eight star points and seven center points (mid-point level 0). This design named face centered CCD mainly due star points of α=1 is in the face of cubic part in the design. The cutting parameters and their levels used in this study as shown in Table 2. The experimental results of face centered CCD developed using MINITAF software consist of 31 trials in coded and actual factors along with output responses of tool flank wear and surface roughness are shown in Table 3.

Table 2 Machining parameters and their levels

			Levels		
Machining parameters	**Units**	**Symbol**	**-1**	**0**	**+1**
Cutting speed	m/min	*V*	50	100	150
Feed rate	mm/rev	*F*	0.05	0.15	0.30
Depth of cut	mm	*D*	0.5	1.0	1.5
Machining time	min	*T*	2	4	6

Response surface method (RSM) was adopted to model the process parameters with the response variables. RSM is the procedure for determining the association among machining parameters and responses and explain influence of process parameters on desired responses. Based on RSM numerical correlation among output responses and input machining parameters expressed as

$$Y = f(V, F, D, T) \qquad \text{(i)}$$

Where *Y* is required output response and *f* is called response function (or response surface). In the investigation procedure, the estimation of *Y* was adapted using the second-order polynomial regression equation; this generally named the quadratic model. The quadratic model of *Y* expressed as given below:

$$Y = a_O + \sum_{i=1}^{4} a_i X_i + \sum_{i=1}^{4} a_{ii} X_i^2 + \sum_{i \pi j}^{4} a_{ij} X_i X_j \qquad \text{(ii)}$$

Where a_O, a_i, a_{ii} and a_{ij} indicate constant, coefficients of linear, quadratic and cross product terms correspondingly. X_i Express cutting variables in the coded form.

The coded variables X_i, i = 1, 2, 3, 4 are obtained from given below transformation relationship equation:

$$X_1 = \frac{V-V_0}{\Delta V} \quad \text{(iii)}$$

$$X_2 = \frac{F-F_0}{\Delta F} \quad \text{(iv)}$$

$$X_3 = \frac{D-D_0}{\Delta D} \quad \text{(v)}$$

$$X_4 = \frac{T-T_0}{\Delta T} \quad \text{(vi)}$$

Table 3 Experimental results for machining of Al/10%SiCp MMC using K10 tool

Sl.	Coded				Actual				Output Responses	
No	X_1	X_2	X_3	X_4	*V*	*F*	*D*	*T*	Y_1 (VB_{max}, mm)	Y_2 (Ra, μm)
1	-1	0	0	0	50	0.15	1.0	4	0.09	2.71
2	1	0	0	0	150	0.15	1.0	4	0.25	2.27
3	0	-1	0	0	100	0.05	1.0	4	0.23	1.92
4	0	1	0	0	100	0.25	1.0	4	0.21	2.95
5	0	0	-1	0	100	0.15	0.5	4	0.18	2.41
6	0	0	1	0	100	0.15	1.5	4	0.22	2.48
7	0	0	0	-1	100	0.15	1.0	2	0.17	2.35
8	0	0	0	1	100	0.15	1.0	6	0.25	2.53
9	-1	1	1	1	50	0.25	1.5	6	0.15	3.25
10	1	-1	-1	-1	150	0.05	0.5	2	0.30	1.59
11	-1	-1	1	1	50	0.05	1.5	6	0.17	2.22
12	1	1	-1	-1	150	0.25	0.5	2	0.22	2.63
13	-1	-1	-1	1	50	0.05	0.5	6	0.15	2.16
14	1	1	1	-1	150	0.25	1.5	2	0.24	2.69
15	-1	-1	-1	-1	50	0.05	0.5	2	0.11	2.01
16	1	1	1	1	150	0.25	1.5	6	0.38	2.93
17	-1	1	-1	1	50	0.25	0.5	6	0.13	3.18
18	1	-1	1	-1	150	0.05	1.5	2	0.31	1.65
19	-1	1	-1	-1	50	0.25	0.5	2	0.08	3.00
20	1	-1	1	1	150	0.05	1.5	6	0.40	1.84
21	-1	1	1	-1	50	0.25	1.5	2	0.09	3.04
22	1	-1	-1	1	150	0.05	0.5	6	0.37	1.79
23	-1	-1	1	-1	50	0.05	1.5	2	0.10	2.06
24	1	1	-1	1	150	0.25	0.5	6	0.34	2.86
25	0	0	0	0	100	0.15	1.0	4	0.20	2.45
26	0	0	0	0	100	0.15	1.0	4	0.19	2.44
27	0	0	0	0	100	0.15	1.0	4	0.21	2.46
28	0	0	0	0	100	0.15	1.0	4	0.20	2.44
29	0	0	0	0	100	0.15	1.0	4	0.21	2.45
30	0	0	0	0	100	0.15	1.0	4	0.20	2.44
31	0	0	0	0	100	0.15	1.0	4	0.20	2.46

Where X_1, X_2, X_3 & X_4 are the corresponding coded values of machining parameters of *V, F, D & T* correspondingly and similarly V_0, F_0, D_0 & T_0 are the values of *V, F, D & T* at zero condition. In further ΔV, ΔF, ΔD & ΔT are the intervals of variation in *V, F, D & T* correspondingly. The tool flank wear (VB_{max}) and average work piece surface roughness (Ra) denoted in Y_{VBmax} and Y_{Ra} correspondingly is evaluated as the output responses. The main objective of using quadratic model is not only examining the whole factor space and also identify area anywhere responses approach it optimum or near optimum value.

4. EMPIRICAL MODELING

4.1. Development of empirical model

The empirical correlation among output responses and cutting variables (V, F, D & T) is given in Equations (vii) & (viii) is obtained by experimental results are shown in Table 3. Tool flank wear (VB_{max}) and surface roughness (Ra) regression coefficient and P-value are shown in Table 4. From the Table 3 observed that P-values of tool flank wear (VBmax) clearly show that linear, square effect of cutting speed, feed rate and machining time, cutting speed interaction with feed rate and machining time, feed rate interaction with machining time are most significant, whereas other effects are not so significant. Similarly, from surface roughness (Ra) P-values from table shows that liner effect of cutting speed, feed rate and machining time, square effect of cutting speed and feed rate, cutting speed interaction with feed rate and machining time, feed rate interaction with machining time are significant whereas other effects are not so significant since the P-values are more than 0.05.

$$VB_{max} = 0.04816 +(0.00351V) -(0.93627f) -(0.05175d) -(0.03224t) -(0.00001V^2) +(2.56533f^2) +(0.02261d^2) +(0.00391t^2) -(0.00150Vf) +(0.00010Vd) +(0.00013Vt) +(0.02500fd) +(0.03125ft) +(0.00375dt) \quad \text{(vii)}$$

$$Ra = 2.38802 -(0.00720V) +(5.34311f) +(0.14201d) +(0.06380t) +(0.00001V^2) -(2.27333f^2) -(0.05093d^2) -(0.00443t^2) +(0.00275Vf) +(0.00005Vd) +(0.00010Vt) +(0.02500fd) +(0.05000ft) +(0.00250dt) \quad \text{(viii)}$$

Table 5 gives the values of estimated standard deviation (S) about the regression line, R^2 statistic and adjusted R^2 statistic. Since the S-value indicates error smaller is good for better model. From the Table 4 observe that S-Value of VBmax is smaller than Ra that indicates VBmax minor amount differed from regression line compared to Ra. Further the R^2 value of Ra higher than VBmax, so the coefficients of empirical equation of Ra more effective than that of VBmax. Furthermore R^2 and adjusted R^2 closeness decides fitness of the model, in this both models R^2 and adjusted R^2 are very close to each other. From the Tables 4 & 5 it can be concluded that developed empirical model clearly model the relationships between the process parameters and output responses VB_{max} and Ra in turning of particulate metal matrix composite.

Table 4 Regression coefficient and P-value of VB_{max} and Ra

Symbol	VB_{max}		Ra	
	Coefficient	P-value	Coefficient	P-value
Constant	0.04816	0.156	2.38802	0.000
v -m/min	0.00351	0.000<	-0.00720	0.000<
f -mm/rev	-0.93627	0.001<	5.34311	0.000<
d -mm	-0.05175	0.397	0.14201	0.087
t -min	-0.03224	0.045<	0.06380	0.005<
v^2	-0.00001	0.003<	0.00001	0.003<
f^2	2.56533	0.002<	-2.27333	0.026<
d^2	0.02261	0.434	-0.05093	0.187
t^2	0.00391	0.041<	-0.00443	0.073
v f	-0.00150	0.018<	0.00275	0.002<
v d	0.00010	0.391	0.00005	0.741
v t	0.00013	0.000<	0.00010	0.016<
f d	0.02500	0.665	0.02500	0.741
f t	0.03125	0.042<	0.05000	0.016<
d t	0.00375	0.205	0.00250	0.511

Table 5 Summary of regression analysis

Responses	**S-value**	**R^2**	**Adjusted R^2**
Flank wear (VB_{max})	0.0113420	99.05%	98.22%
Surface roughness (Ra)	0.0148850	99.94%	99.88%

5. RESULTS AND DISCUSSION

5.1. Influence of cutting parameters on flank wear (VB_{max})

During machining of heterogeneous material like Al/SiC MMC cutting tools exposed higher wear and cyclic load mainly due to presence of higher hardness SiC particulate in the Al matrix. The developed empirical equation based on experimental using RSM given the Equation- (vii). In the study examine the effect of four different cutting parameters on output response of tool flank wear (VB_{max}). The contour plots are developed for different cutting combinations; numbers in the pols denote the amount tool flank wear in mm.

The study of the influence of process parameters on the generation of flank wear and tool wear characteristics creates a basis for understanding the machinability and reducing or preventing tool wear in order to ensure the integrity of the product. This is more important in the case of advanced materials like SiCp reinforced aluminium MMC. Fig.1 illustrates the influence of cutting speed and feed rate on $VB_{max.}$ It is seen that cutting speed increases from lower value 50 m/min to higher values of 150 m/min the flank wear also increases. At lower cutting speed of 50 m/min, possible formation of unstable larger BUE protects the cutting wedge from the propagation of wear, but this is eliminated when the cutting speed is increased to 150 m/min, thus resulting in higher tool wear at higher cutting speed. Also, when the cutting speed is increased, the cutting temperature increases and there is linear relationship between the two. Work done in machining is converted into heat that eventually leads to increase in cutting temperature, which not only reduces the BUE but also softens the tool thus leading to higher tool wear [18]. This is further accelerated by the interaction between the harder SiC particles and carbide cutting tool. Fig.2 shows the BUE formation at low and higher cutting speed.

From Fig.1 with increase in feed rate, a decrease in flank wear (VB_{max}) is seen up to around 0.20 mm/rev of feed rate. At lower feed rates the contact time between tool - work and tool - chip is more which leads to higher abrasion of SiCp present in the work on cutting edge thus also leading to greater heat concentration and thus resulting in more tool wear [19]. However, with increase in feed rate beyond 0.20 mm/rev increase in flank wear (VB_{max}) is seen. With higher feed rate, consequent larger cross- sectional area of the un-deformed chip and resistance to chip deformation leads to higher change in the geometry of tool wedge. The changes in the geometry further lead to thermal induced tool wear thus resulting in higher flank wear.

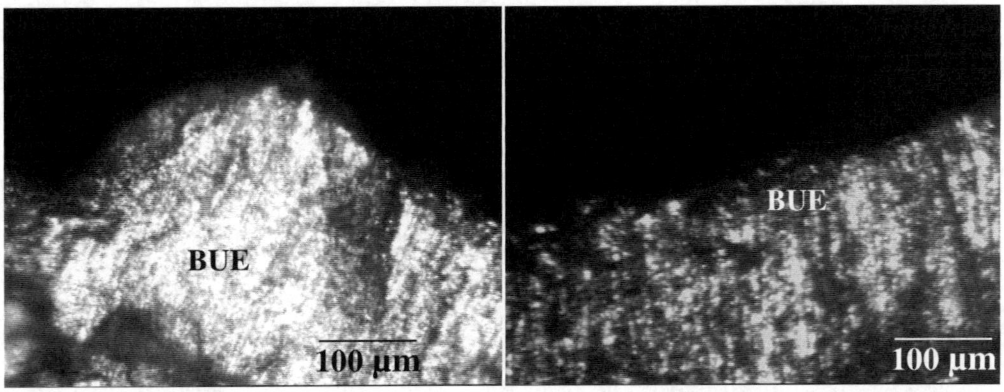

(a) *V*-50, *f*-0.05, *d*- 0.5, and *t*-2 **(b) *V*-150, *f*-0.05, *d*- 0.5 and *t*-2**

Fig.1 BUE formation during machining of Al/10%SiCp MMC using K10 tool

Fig.3 illustrates the influence of machining time and depth of cut on flank wear. The flank wear generally increases with increase in machining time, due to degradation of cutting tool and alteration in the form of tool wedge [20]. The effect of depth of cut on flank wear is less seen from the Figure. This result correlates results of P value given in Table 4 wherein the depth of cut is observed as the least influence factor compared to cutting speed, feed rate and machining time on tool flank wear in machining of Al/SiC MMC [21]. But depth cut increased in to above 1 mm, the higher tool flank wear observed this mainly due increases in work-tool contact area, normal cutting force and friction its leads to rises interfacial cutting temperature. The higher interfacial temperature further leads to softening of work material its increases tool flank wear.

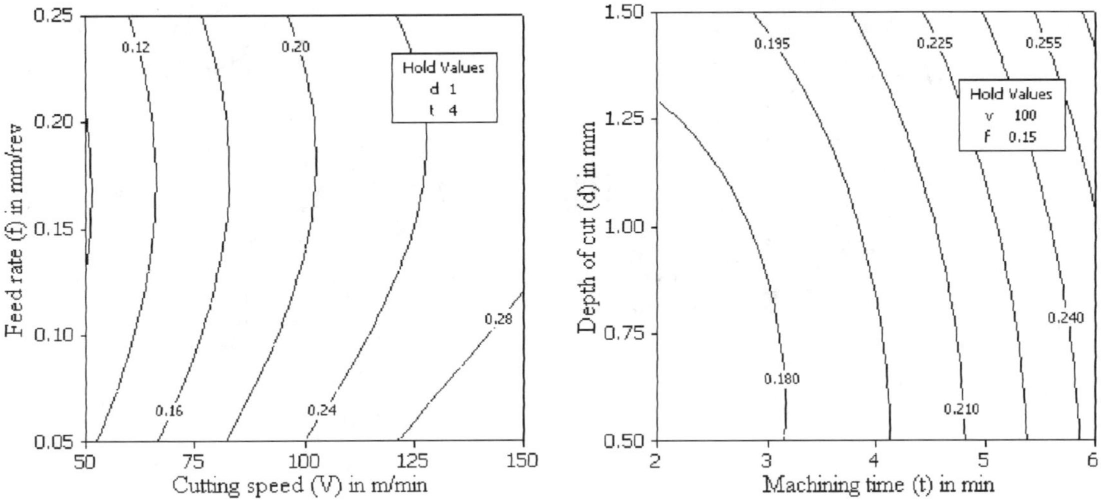

IOP Conf. Series: Materials Science and Engineering **961** (2020) 012014 doi:10.1088/1757-899X/961/1/012014

Fig.2 Influence of cutting speed (*V*) and feed rate (*F*) on Tool flank wear

Fig. 3 Influence of Depth of cut (*D*) and Machining time (*T*) on Tool flank wear

Several types of wear mechanisms can influence the tool geometry. The worn cutting edges of K10 grade cemented carbide cutting tool observed under scanning electron microscope (SEM) while turning at low cutting speed of 50 m/min is shown in Fig.4. From the Fig.4 distinct abrasive wear grooves can be seen at the work flow direction of primary cutting edge of flank, due to two bodies and three bodies abrasive action of hard reinforced SiC particles [21]. The two body abrasive wear is produced through the roughing action of SiC particulates presents in Al matrix moving at higher cutting speed against the cutting tool. The three body abrasive wear is caused by the released hard SiC particles, entrapped between the tool and the work piece [22].

The wear of cutting tool at a cutting speed of 150 m/min is shown in Fig.5. At higher cutting speed, in addition to abrasive wear, triangular wear on nose region and wear in the chip flow direction on the rake face is also seen. This is mainly due to higher sliding of flow of chips with hard particles at higher cutting speed. This sliding of chips combined adhesion–abrasion of work material on the rake face away from the cutting edge could have resulted in degradation of tool over the flank portion [23].

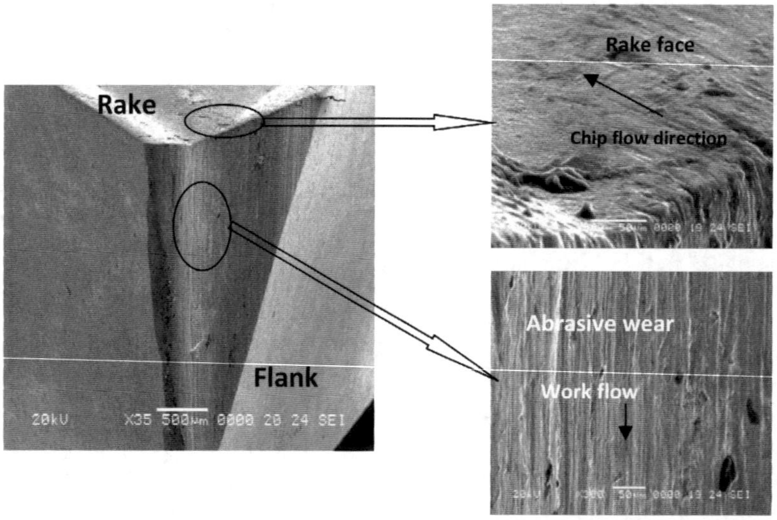

(a) Tool flank wear **(b) Magnified circle area**

Fig.4 SEM of Cutting tool (K10) at *V*= 50m/min, *F*=0.05mm/rev, *D*=1.5mm and *T*= 6 min

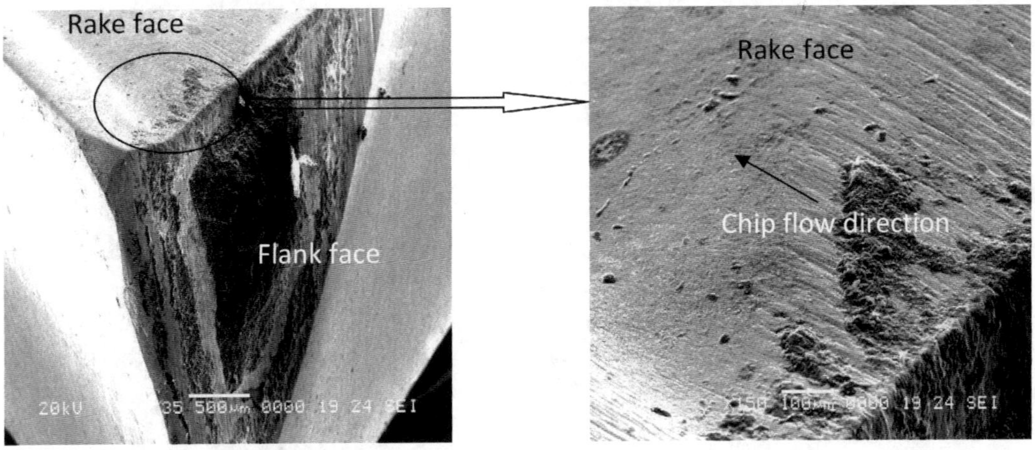

(a) Tool flank wear **(b) Magnified circle area**

Fig.5 SEM of Cutting tool (K10) at *V*=150 m/min, *F*=0.05 mm/rev, *D*=1.5 mm and *T*=6 min

The abrasive wear in work flow of flank face and chip flow direction rack face, combined adhesion-abrasive wear observed during 100 m/min is shown in Fig.6. The main wear mechanism observed K10 tool during machining Al/10%SiCp MMC is purely abrasion at low cutting speed of 50 m/min, combination of abrasion and adhesion observed at higher cutting speed [24]. The worn cutting edges observed from SEM also reveal the triangular and grooved wear in the cemented carbide cutting tools.

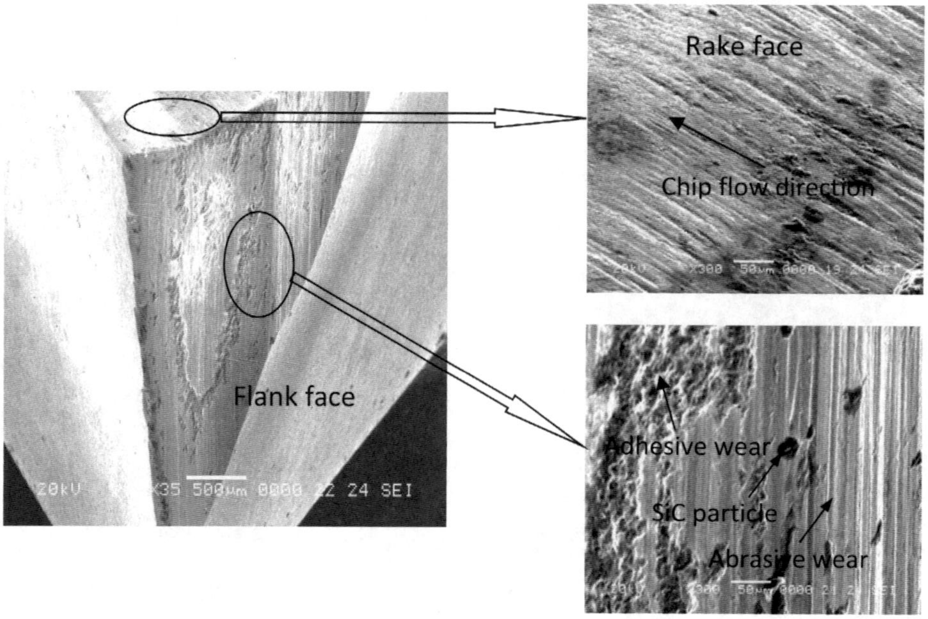

(a) Tool flank wear **(b) Magnified circle area**

Fig.6 SEM of Cutting tool (K10) at *V*- 100 m/min, *F* -0.05 mm/rev, *D*- 1.0 mm and *T*- 4 min

5.2. Influence of cutting parameters on surface roughness (Ra)

The Fig.7 shows effect of cutting speed (V) and feed rate (F) on the out response of surface roughness (Ra). From Fig.7 observed that the surface roughness (Ra) increases as the feed rate increases. Higher feed rates increase the temperature in cutting zone causing decrease in bonding effect between SiCp and Al matrix, thus leading to pull out of SiC particles from the softened aluminum matrix leaving a cavity on the machined surface resulting in poor surface finish [122]. We know that at higher feed rate the cusp height is more which also leads to higher values of surface roughness. From Fig. 2 at low cutting speed, larger size unstable BUE formed this break the chips frequently this leads the poor surface roughness. But increases cutting speed higher interfacial temperature eliminate unstable BUE this reduce the chip fracture leads to good surface finish [23]. The finest required surface morphology of work piece is attained at lesser feed rate and higher cutting speed cutting combinations during machining Al/10%SiCp MMC using K10 tool.

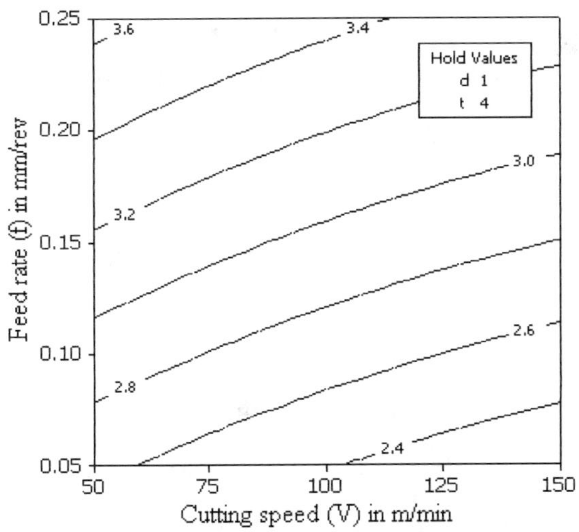

Fig.7 Influence of Cutting speed (*V*) and Feed rate (*F*) on Ra

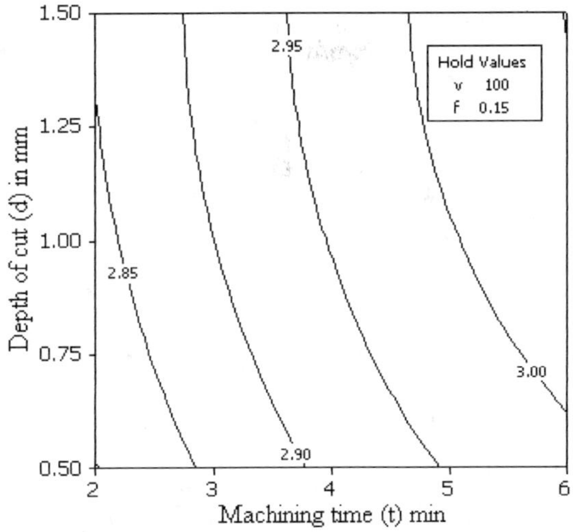

Fig.8 Influence of Machining time (*T*) and Depth of cut (*D*) on Ra

The effect of depth of cut and machining time on the surface roughness is shown in Fig.8. The depth of cut is lower influencing machining parameter on surface roughness compared to cutting speed, feed rate and machining time [21], which is seen from the observed 'P' values given in Table 4. From the illustration it is observed that the surface roughness increases with machining time because tool geometry gets altered.

6. ANALYSIS FOR OPTIMIZATION OF THE RESPONSES

Apart from the study of interaction between process parameters and their effect on desired response variables, it is necessary to optimize the process parameters for the multiple responses of tool flank wear and average surface roughness. One useful approach to optimize the multi-objective simultaneous optimization method procedure developed by researches Derringer and Suich [13]. Their procedure introduces the concept of desirability functions. The general approach is to first convert each response Y into an individual desirability function d_i that varies over the range.

$$0 \leq d_i \leq 1 \qquad \text{(ix)}$$

When the response Y is at its goal or target then $d_i = 1$ and if the response is outside an acceptable region then $d_i = 0$. The weight of the desirability function for each response defines its shape. For each response, one can select a weight (r_i) to emphasize or de-emphasize the target. Finally the individual desirability functions are combined to provide a measure of the composite or overall desirability of the multi-response system [10]. This measure of composite desirability is the weighted geometric mean of the individual desirability of the responses. The optimal operating conditions can be determined by maximizing the composite desirability [11]. In the present investigation, the response parameters are chosen to maximize the overall desirability as follows: $D = (d_1^{i1}\ d_2^{i2})^{1/(i1+i2)}$ (x)

Where d_1 and d_2 denote VBmax and Ra desirability functions respectively and i1 and i2 is parameters of transformed responses of d_1 and d_2. Generally the compact gradient algorithm has various starting points it is used for maximize/minimize the compound desirability to identify the optimal input parameter combination. Many standard statistical software tools like Minitab, Design expert, etc. are used to optimize the responses. In this study Minitab is used to optimize the response parameters.

The Fig. 11 shows the single desirability of optimization plot for minimizing tool flank wear. The upper and target values for optimization is chosen from experimental results as showed in Table 3 as 0.39 mm and 0.07 correspondingly. The optimized cutting parameters for attaining minimum tool flank wear of 0.07 mm from optimization plot as follows Cutting speed (*V*) = 50 m/min, Feed rate (*F*) = 0.18 mm/rev, Depth of cut (*D*) = 1.0 mm and Machining time (*T*) = 4 min. All the machining parameters are within range, the desirability of optimization is 100%.

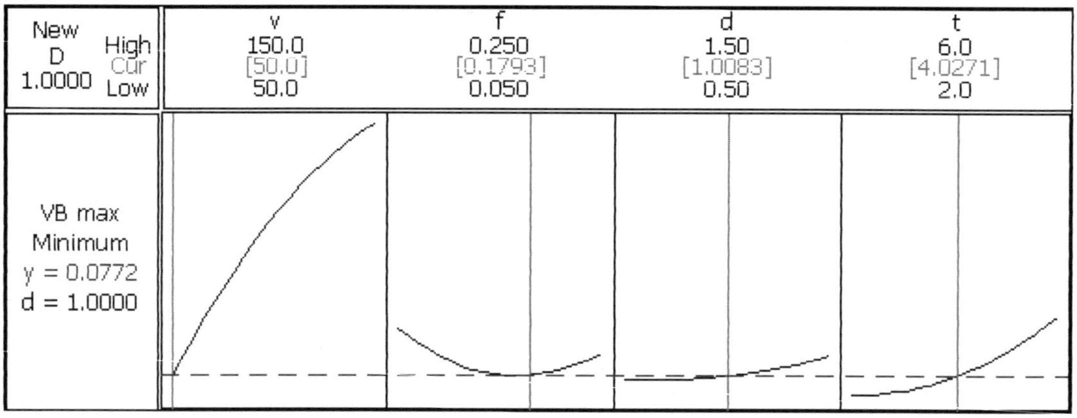

Fig.9 Optimum results for minimum flank wear (VB_{max})

Similarly optimization plot for minimum surface roughness (Ra) is shown in Figure 12. From Table 3 the target and upper values chosen for optimization are 1.59 μm and 3.18 μm correspondingly. The optimized cutting parameters for attaining minimum surface roughness of 1.59 μm from optimization plot as follows Cutting speed (*V*) = 150 m/min, Feed rate (F) = 0.05 mm/rev, Depth of cut (D) = 0.6 mm and Machining time (T) = 2 min. All the machining parameters are within range, the desirability of optimization is 100%.

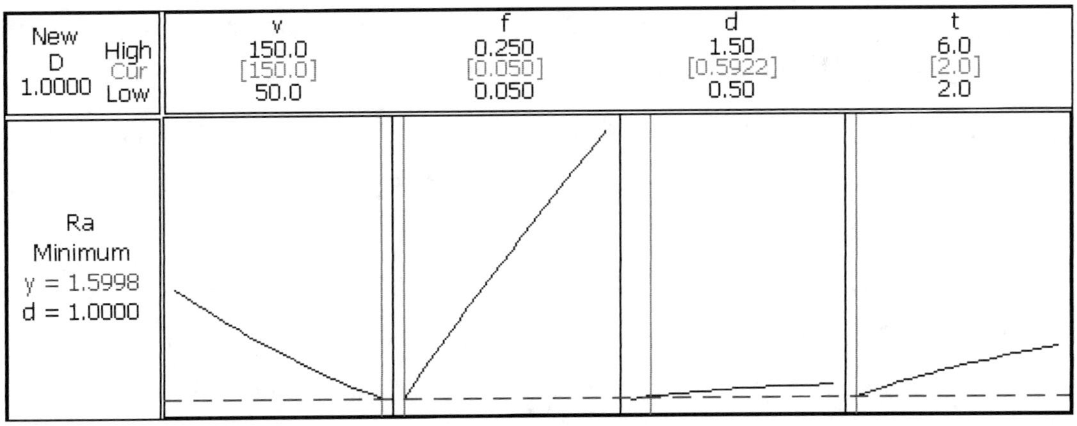

Fig.10 Optimum results for minimum surface roughness (Ra)

The goal was to minimize VB_{max} and Ra while both are considered at a time. As the desirability is close to one, it can be concluded that the parameters are within their working range. Optimization plot for the both responses is shown in Fig.11. The optimum values from the plot are VB_{max} = 0.18 mm and Ra = 2.36 µm and the corresponding parameters of Cutting speed (V) = 80 m/min, Feed rate (F) = 0.05 mm/rev, Depth of cut (D) = 0.50 mm and Machining time (T) = 2 min respectively.

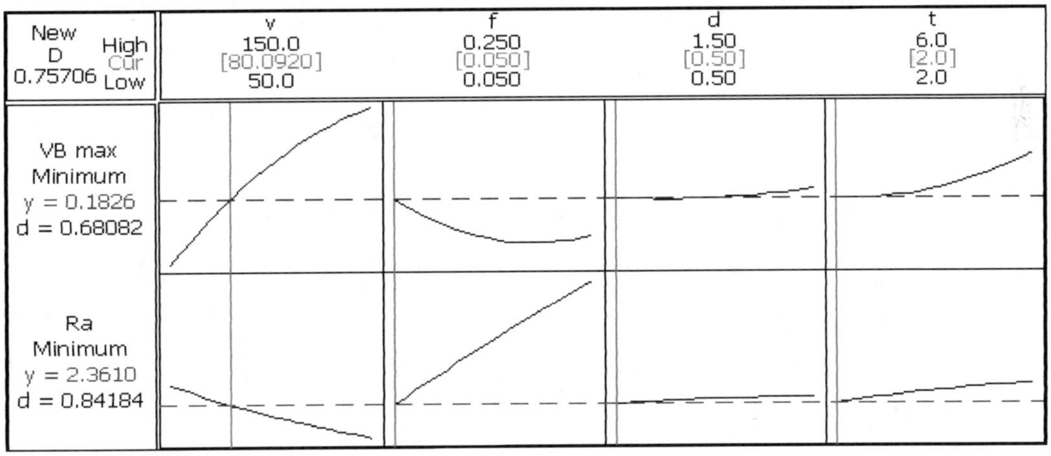

Fig.11 Optimal chart of minimum VB_{max} and Ra obtained through RSM for Al/10%SiCp MMC using K10 tool

7. CONCLUSIONS

The response surface methodology based disability approach optimization and effect of machining parameters on tool flank wear and surface roughness during machining composite machining are analyzed in this study. From analyze, the following conclusions are made.

1. From the empirical models found that cutting speed and feed rate more dominant when compared to other parameters. The suggested numerical models for tool flank wear and surface roughness is found to be suitable and used for predict the parameters within this experimental range.
2. The BUE formation adversely affects tool flank wear especially at low cutting speeds whereas higher cutting speeds and feed rates work material thermal softening reduce BUE formation.Surface topographies of the tool indicate that the main wear mechanism of cemented carbide tool is abrasive and adhesive wear.
3. The BUE formations at low cutting speeds produce poor surface finish. The better surface roughness achieved at lower feed rats and high cutting speed combination. Depth of cut increases the increase in surface roughness.
4. The optimized combination cutting parameter found from RSM based desirability approach such as Cutting speed (*V*) = 80 m/min, Feed rate (*F*) = 0.05 mm/rev, Depth of cut (*D*) = 0.50 mm and Machining time (*T*) = 2 to attain the minimum Tool flank wear (VB_{max}) and surface roughness (Ra) of 0.18 mm and 2.36 μm respectively.

REFERENCES

[1] El-Gallab M and Sklad M 1998 J Mater Process Technol. 83: 277–285.

[2] Uday A Dabade, Suhas S. Joshi, Balasubramaniam R and Bhanuprasad V V 2007 J Mater Process Technol.192–193: 166–174.

[3] Pramanik A, Zhang L C and Arsecularatne J A 2008 Int. J. Mach. Tools Manuf. 48: 1613–1625.

[4] Kannan S and Kishawy H A 2008 J Mater Process Technol. 198: 399–406.

[5] Seeman M, Kanagarajan D, Seetharaman R and Sivaraj P 2017 http://ijamtes.org/ VII (III): 137-151.

[6] Muthukrishnan N, Murugan M and Prahlada Rao K 2008 Int. J. Adv. Manuf. Technol. 39: 211–218.

[7] 62. Palanikumar K and Karthikeyan R 2007 Mater. Des. 28: 1584–1591.

[8] Seeman M, Ganesan G, Karthikeyan R and Velayudham A 2010 Int. J. Adv. Manuf. Technol. 48: 613-624.

[9] Tomac N, Tonnessen T and Rasch F O 1992 Ann CIRP. 41 (1): 55-58.

[10] Looney L A, Monaghan J M, O'Reilly P and Taplin D M R 1992 J Mater Process Technol. 33: 453-468.

[11] Lin J T, Bhattacharyya D and Lane C 1995 Wear 181-183: 883-888.

[12] Quigley O, Monaghan J and Reilly P 1994 J Mater Process Technol. 43: 21-36.

[13] Sahin Y 2005 Surf Coat Tech. 199: 112– 117.

[14] Muthukrishnan N, Murugan M, Prahlada Rao K 2008 Int. J. Adv. Manuf. Technol. 38: 447–454.

[15] Paulo Davim J 2007 Mater. Des. 28: 2684-2687.

[16] Paulo Davim J 2002 J Mater Process Technol. 128: 100–105.

[17] M. Seeman, D. Kanagarajan, P. Sivaraj, R. Seetharaman and A. Devaraju 2019 IOP conf. ser. Mater. Sci. Eng. 574; 012008

[18] Tay A A O and Lee K H J Mater Process Technol. 29: 47–62.

[19] Pathak J P and Tiwari S N 1995 Indian Foundry J. 41 (1): 1-8.

[20] Manna A and Bhattacharayya B 2005 Int. J. Adv. Manuf. Technol. 25: 850-856.

[21] Uday A Dabade, Dilip Dapkekar and Suhas S Joshi 2009 J. Mach. Tools Manuf. 49: 690 – 700.

[22] Xiaoping Li and Seah W K H 2001 Wear 247: 161–171.

[23] Kilic kap E, Cakir O, Aksoy M and Inan A 2005 J Mater Process Technol. 164–165: 862–867.

[24] Zhang Z F, Zhang L C and Mai Y W 1995 J. Mater. Sci.30 (8):1967–1971.

Investigation on the Mechanical and Morphological Properties of Red banana/Ramie Fiber vinyl ester composites

G.Sai Krishnan[1], G.Shanmugasundar[2], M.Vanitha[3], S.Srinivasan[1], G.Suresh[1]

[1]Department of Mechanical Engineering, Rajalakshmi Institute of Technology
[2]Department of Mechanical Engineering, Sri Sairam Institute of Technology
[3]Department of Chemistry, Sri Sairam Engineering College
*Corresponding author: g.saikrishnan@gmail.com

Abstract.
The need of green materials for various applications is in high demand. The bio fibers play a very important position to achieve eco friendliness and cost effectiveness. The aim is to analyse the effect of hybrid mat performance. In this work banana and ramie fibers were taken into consideration, With vinyl ester composites. Three different combinations have been developed in laminating the composites at 90 degree orientation. ASTM standards were used to examine the mechanical performances of the composites. The worn out analysis have been carried out by using scanning electronic microscopy. In this work two single layer fibers of ramie and banana have been taken into account and the third composite is developed with hybridisation of red banana and ramie fibers. On comparing all the results obtained the hybrid double layer of red banana and ramie fibers possessed good results in tensile /flexural property. The impact properties of ramie fiber were marginally higher than red banana fibers.

Keywords: Red Banana Fiber, Ramie Fibers, Vinyl Ester Composite, Scanning Electronic Microscopy (SEM).

1.Introduction

Need of green resources is in high demand because of its versatile applications.It is in increasing because of its ease manufacturing and cost effectiveness. The safe of handling characteristics of such materials in functioning ambience in industries in considered being effective. The work, exploitation of such natural fibers has attracted the researchers. The natural fibers are identified based on their geographical location, cost effectiveness and eco friendly. These fibeorus materials are used as an key in ingredient in various applications such as automobile, household and aero space [1-6]. These fibers are used in various forms with different orientation; It is done in random orientation, Continuous orientation and in woven form. Out of various fibers used woven fibers were identified as an attractive one due to its enhanced mechanical strengths. Various fibers are used in various applications such as reinforcements. The different parts of plants are utilised in various applications. Geographical location is considered to be vital in utilisation of fibers for composite applications. The crystalline and amorphous contents are considered to be effective in manufacturing the composites [7].Less the amorphous content more the strength of the fiber. The various fibers with different stacking sequences and different orientation have been tried by various researchers. Vijay[8] investigated the mechanical and morphological properties of Areca Sheath and palm leaf sheath and concluded that the areca fibers as a skin and palm as a core possessed better mechanical strengths. Sai Krishnan Et al [9] analysed the performance of bamboo and jute fibers and concluded that the bamboo at the outer layer and jut at the inner layer had better mechanical properties. Jothibasu et al. [10] deliberates the performance of areca sheath/ jute/glass fiber-hybrid composites and he concluded that the hybrid glass/jute/arecasheath/jute fiber composites proved improved tensile, flexural, compression, and shear properties when compared to other composites. Chaudhary et al. [11] investigated the performance of hybrid reinforced polymeric composites. Out of all the developed composites the hybrid composites had better results. Ganesh babu Et al investigated the chemical treatment of cypress pangorei and developed in form of brake pad and he concluded that chemical treated fiber had impact on overall tribological results[12].Based on this various outcome obtained, the lack of hybrisation is observed. In this work a combination of red banana and ramie fibers were utilised in the woven form and work is carried out. Mechanical strengths of hybrid composites were studied and fractured specimen were analysed by using SEM [13-15].

2. Materials and Methods (Preparation of Composites)

Red banana and Ramie fibers in the mat form were procured from jeeva natural fibers Chennai. They were procured in the form of mat and usually grow in various Asian countries. In general banana fibers were extracted from the stem part of banana. In this work vinyl ester resin were used as a matrix and the chemical composition of the used fibers is shown in the table 1 below. Methyl Ethyl ketone peroxide and cobalt naphthaate is used in this work as a catalyst and accelerator. Compression moulding technique was used in this work. They were kept at 90 degree orientation with mold dimensions of 200x200x4 mm^3. All the tests were carried out as per ASTM standards. The single fiber tests for the red banana and ramie fibers were done by using zwickroell- zwickiline material testing machine. The Three composites were developed and designated as C1-Single layer red banana fiber mat(SLRBFM), C2-Single layer ramie fiber mat(SLRFM), C3-Hybrid composites of red banana fiber and ramie fiber mat(HRBRFM).The description of composites is shown in table 1and the images of composite in mat form is shown in figure 1 below.

Table 1. Fiber layers and Acronyms for different layers of composites

Sl.no	Designation Of Composites	Fiber Layers	Acronyms
1	C1	Single layer red banana fiber mat	SLRBFM
2	C2	Single layer ramie fiber mat	SLRFM
3	C3	Hybrid composites of red banana fiber and ramie fiber mat	HRBRFM

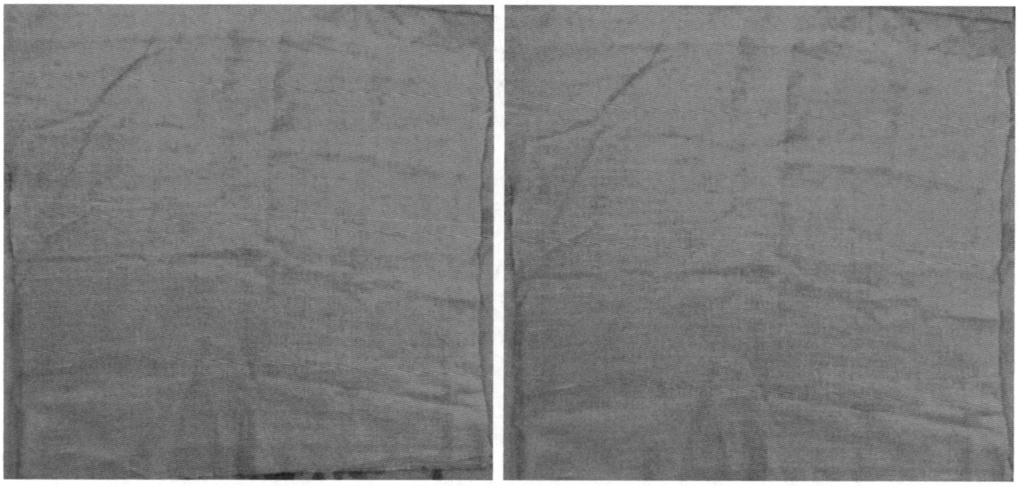

Figure1:Images of Redbanana and Ramie fiber in mat form

3. Results and Discussion

3.1 Tensile Properties

Tensile property for evaluated Red banana/Ramie Fiber is shown below in figure 3. Single layer mat of SLRBFM and SLRFM have been analysed. Along with this Hybrid fiber properties were also evaluated. In a single layer fiber mat orientation, tensile strength for the red banana fibers was considerable higher than that of the ramie fiber. Out of all the composites the three composites developed the hybrid composites possessed better results. It is observed that the hybrid composites possessed better results at 90 degree orientation. This is mainly attributed owing to its high stiffness property of the fiber and load transfer between the fiber and the matrix. The single fiber strength of red banana had higher strength owing to its transformation in load between the fiber and the medium. Same observations was seen in the previous literatures done before [16-18].

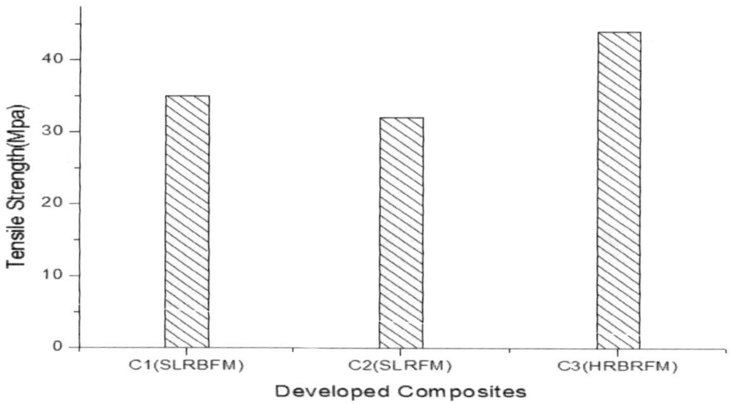

Figure2: Developed Composites (Examination of Tensile Strength)

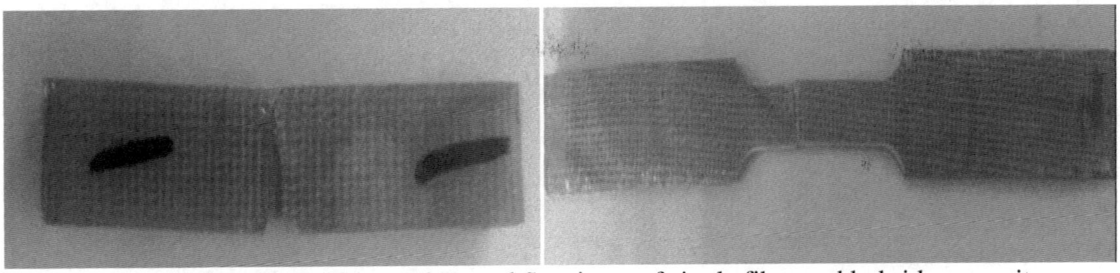

Figure 3:Tensile and Flexural Tested Specimen of single fiber and hybrid composites

3.2 Flexural Properties

The Flexural Strength of the developed composites is analysed and the details are given below. Flexural specimens of red banana fiber mat composites and ramie fiber mat is analysed and it is shown in the figure 3.Single layer mat orientation reinforcements for both C1-Single layer red banana fiber mat (SLRBFM), C2-Single layer ramie fiber mat(SLRFM) was carried out at 90 degree orientation. The strength of both the composites were varied in between 28 MPa to 42 Mpa. Among all the developed composites the double layer composites with hybridisation of both red banana and ramie fibers had better results. This may be attributes due to its orientation and interfacial bonding between the fiber and the matrix. The same thing is witnessed in various literature work done before[19].

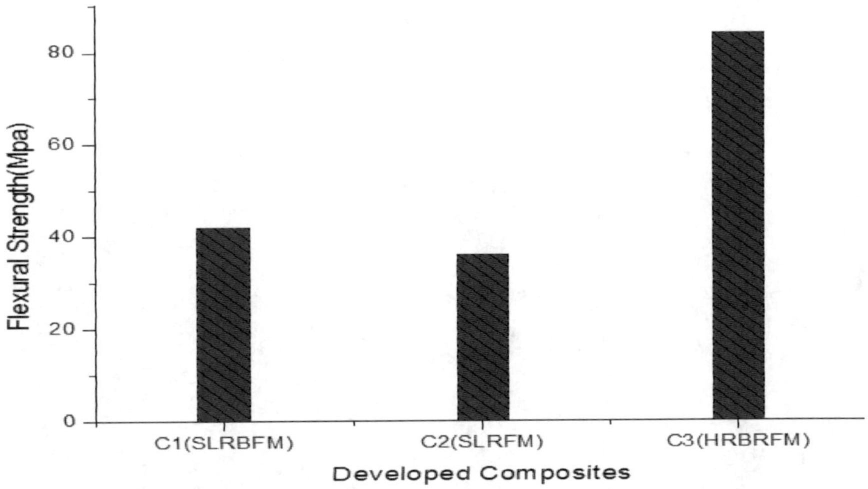

Figure4: Developed Composites (Examination of Flexural Strength)

3.3 Impact Strength

The Energy absorbed during machining operations is carried out with the help of impact strength. The hybrid composites at 90 degree orientation had better results. It is identified based on the results that the Single layer red banana fiber mat (SLRBFM) had less value than that of the Single layer ramie fiber mat. Out of all the composites the Hybrid composites of red banana fiber and ramie fiber mat (HRBRFM) composites possessed better results. The enhancement in the hybrid layers is mainly due to the number of layers of mat present. The Single layer ramie fiber mat had better results due to its high energy absorption capacity. The single layer red banana fiber absorbed less strength due to which it impact strength gets decreased [20].

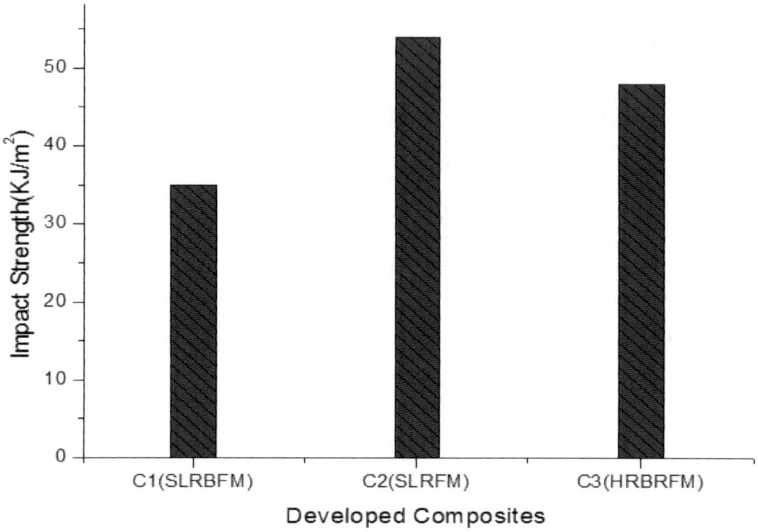

Figure 5: Developed Composites (Examination of Impact Strength)

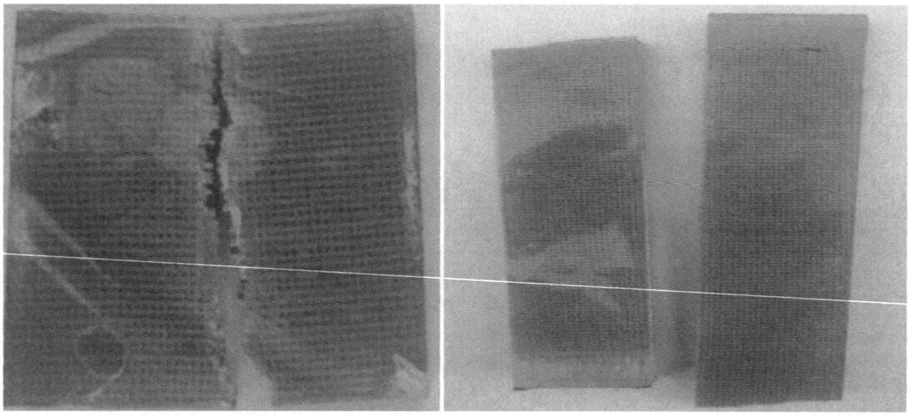

Figure 6:Impact Tested Specimen Sample

4. Conclusion

In present research, Red banana fibers and Ramie fibers were fabricated and the various Properties such as tensile, Flexural and impact properties have analysed. The 90 degree orientation of the fibers enhanced the results of all the composites. Based on the various results obtained it can be concluded that the highest tensile and flexural strength is obtained for Hybrid composites of red banana fiber and ramie fiber mat(HRBRFM) at 90 degree orientation. It is mainly attributed because of its hybridisation technique between the red banana and ramie fibers. The impact strength is best for the rmie fiber mat composites because of its energy absorption capacity. Depending on these experimental results, one can easily produce low cost environment friendly high strength polymer matrix composites for automobile industry as well as domestic purpose.

5.References

[1] Jafrey Daniel D, Sai Krishnan G and Velmurugan PInvestigation on the characteristics of bamboo/jute reinforced hybrid epoxy polymer composites Mater. Res. Express 6 105346

[2] Sai Krishnan G, Jayakumari L S, Babu L G and Suresh G (2019) Investigation on the physical, mechanical and tribological properties of areca sheath fibers for brake pad applications Mater. Res. Express 6 085109

[3] C.Sivakandhan, Ganesh Babu Loganathan (2020) Material characterization and unconventional machining on synthesized Niobium metal matrix Mater. Res. Express 7 015018

[4] James J D, Manoharan S, Saikrishnan G and Arjun S (2019) Influence of bagasse/sisal fibre stacking sequence on the mechanical characteristics of hybrid-epoxy compositesJ. Nat. Fibers 2 1–11

[5] Saikrishnan G, Jayakumari L S, Vijay R and Lenin Singaravelu D (2019) Influence of iron–aluminum alloy on the tribological performance of non-asbestos brake friction materials–a solution for copper replacement Industrial Lubrication and Tribology 00 1–13

[6] Manoharan S, Shihab A I, Alemdar A S A, Ganesh Babu L, Vijay R and Lenin Singaravelu D 2019 Influence of recycled basalt-aramid fibres integration on the mechanical and thermal properties of brake friction composites Material Research Express 6 115310

[7] G Sai Krishnan, S Kumar (2020) Role of metal composite alloys in non-asbestos brake friction materials-A solution for copper replacement Materials Today: Proceedings 2020/3/31

[8] Vijay R and Singaravelu DL(2016). Experimental investigation on the mechanical properties of Cyperuspangorei fibers and jute fiber-based natural fiber composites. International Journal of PolymerAnalysisi and Characterization 2016; 21: 617–627

[9] Saikrishnan Ganesh Influence of Stacking Sequence on the Mechanical and Water Absorption Characteristics of Areca Sheath-palm Leaf Sheath Fibers Reinforced Epoxy Composites J Nat Fib 2020; 07: 2020

[10] Jothibasu S, Mohanamurugan S and Vijay R. Investigation on the mechanical behavior of areca sheath fibers/jute fibers/glass fabrics reinforced hybrid composite for lightweight applications. J Ind Tex 2020; 49: 1036–1060. DOI: 10.1177/1528083718804207

[11] Chaudhary V, Bajpai PK and Maheshwari S. Studies on mechanical and morphological characterization of developed jute/hemp/flax reinforced hybrid composites for structural applications. J Nat Fib 2018; 15: 80–97

[12] Ganesh Babu, L. (2019). Influence of benzoyl chloride treatment on the tribological characteristics of Cyperus pangorei fibers based non-asbestos brake friction composites. *Materials Research Express*, *7*(1), 015303

[13] G.Sai Krishnan Babu L G and Pradhan R (2020) Study on tribological properties of palm kernel fiber for brake pad applications Materials Research Express Mater. Res. Express 7 015102

[14] Nongman. A. Fa, A. Baharina, and A.AbuBakarb ,—The Effect of Banana Leaves Lamination on the Mechanical Properties of Particle Board Panel‖ 5th International Conference on Recent Advances in Materials, Minerals and Environment (RAMM) & 2nd International Postgraduate Conference on Materials, Mineral and Polymer (MAMIP), 4-6 August 2015

[15] Ramnath, B. V., Kokan, S. J., Raja, R. N., Sathyanarayanan, R., Elanchezhian, C., Prasad, A. R., and Manickavasagam, V. M. (2013) Evaluation of mechanical properties of abaca–jute–glass fibre reinforced epoxy composite. Materials & Design, 51: 357-366.

[16] Vinod A, Vijay R and Singaravelu DL. (2018) Thermomechanical characterization of Calotropis gigantea stem powder-filled jute fiber-reinforced epoxy composites. J Nat Fibers Journal of Natural Fibers; 15: 648–657.

[17] G Sai Krishnan(2020) Investigation on the physical and mechanical characteristics of demostachya bipinnata reinforced with polyester composites Materials Today: Proceedings 2020/4/

[18] Tej Singh Sachin Tejyan Amar Patnaik(2019) Fabrication of waste bagasse fiber-reinforced epoxy composites: Study of physical, mechanical, and erosion propertiespolymer composites—2019

[19] Maha M. Ibrahima, Alain Dufresneb, Waleed K. ElZawawya,*, Foster A. Agblevorc —Banana fibres and micro fibrils as lignocellulosic reinforcements in polymer composites‖ Article in Carbohydrate Polymers July 201

[20] Amira,*, Kamal AriffZainalAbidinb,FaizzatyBintiMdShiric —Effects of Fibre Configuration on Mechanical Properties of Banana Fibre/PP/MAPP Natural Fibre Reinforced Polymer Composite‖ Advances in Material & Processing Technologies Conference

IOP Conf. Series: Materials Science and Engineering **961** (2020) 012016 doi:10.1088/1757-899X/961/1/012016

Investigation on the Physical/Mechanical Properties of NAO brake friction Composites by using Kenaf fiber

P.Ramesh[1],G.Sai Krishnan[2],R.Ragu[3],G.Suresh[4],S.Srinivasan[5]

1Department of Mechanical Engineering, Academy of Maritime Education and Training
[2]Department of Mechanical Engineering, Rajalakshmi Institute of Technology
*Corresponding author: g.saikrishnan@gmail.com

Abstract.

Natural fibers utilization for various applications is in increasing trend owing to its eco friendliness and ease availability. In this work the effect of kenaf fibers as key ingredient is studied for brake pad applications. The Consequences of kenaf fiber as reinforcement is studied for physical and mechanical properties. Kenaf fibers have been added in 3 different weight percentages from 0 wt% to 15 wt%. The developed composites were evaluated as per industrial standards. Properties such as density, ash content,hardness,water absorption and compressibility have been studied in this work. The brake pad samples were prepared and subjected for testing for various physical and mechanical properties. All the tests have been carried out by standard industrial procedure. Based on the results it can be clinched that the properties of hardness, density and content of ash density, decreased to certain extent and other properties such as compressibility, porosity and water absorption capacity increased with fiber loading of kenaf fiber. It can be concluded that the fiber loading has impact on the performance of fibers.

Keywords: Natural Fibers, Kenaf Fibers, Mechanical properties, Fiber Loading, Brake Pad

1.Introduction

In Recent years, attention towards natural fibers increased drastically. The natural fibers have been utilised in various industries for its versatile applications and its cost effectiveness[1-5].Polymeric materials are used as an key friction ingredient material in various automotive applications. For a friction material there are some desired quality to possess such as stable coefficient of friction, resistance to abrasion and wear, high recovery and least fade value.Fricitonal materials should also have less noise and it should not vibrate in an extensive environment. In order achieve all these qualities in frictional material applications around 10-12 ingredients are added[6-7]. It constitutes of binder/filler/frictional modifier and reinforcement fibers. Out of all of these ingredients the roles played by fibers are important. The fibers with the desired combinations is vital is examining the performance of brake pad composites.

Numerous works have been carried out by using various different fibers such as palmkernel,areca,ramie,banana,glass,carbon,basalt Etc. These fibrous materials have been used as an reinforcement to maintain stable friction. Out of all these fibers some fibers had some fibers are expensive and repeatability was seemed to be difficult, and the other fibers such as copper and asbestos may cause serious health issues and may spoil aquatic life badly. so ban of copper by 2021 is already in real world scenario. The various types of fibers were used by various researchers with some modifications and without modifications[8-10]. MaY et al analysed the performance by using two different fibers in his two different research works. He analysed the performance by various fiber loading weight percentages. He varied the weight percentages of the fibers. Initially he carried put his work by using rattan as a reinforcement materials at different loading conditions. He carried out in 4 different laoding conditions such as 2.5,5,7.5 and 10 percentages. He concluded that the 5 weight percentages of rattan fibers possessed good frictional values[11].Likewise, Sai Krishnan Et al analysed the performance of areca fibers as a key varying material and he concluded areca sheath fibers at 5 weight percentages had good fade and recovery. He also tried frictional performances with palm kernel fibers and concluded palm fibers with additives enhanced the frictional coefficient and wear resistance[12-13].Tej singh et al examined the tribological performance by using ramie fibers and concluded that the maximum frictional coefficient was achieved at 5 weight percentage loading and had excellent tribological properties[14].Similarly various fibrous reinforcement materials have been tried and excellent out come was revealed by using this materials[15-17].Based on the various literatures done it was evident that the fibrous materials had greater impact on frictional performances.so it can be used as an potential replacement. In this research work, kenaf fibers were identified owing to its excellent mechanical properties and eco friendliness. It was selected and used as an reinforcement material in polymeric applications[18-20].Though it had various advantages its potential utilization in frictional materials applications have not been explored so far.Owing to this it was used as an reinforcements and added as a key ingredient in developing the brake pad. The composites brake pad were developed by varying in four different weight percentages from 0 to 15 weight percentages. The basic properties were evaluated in this work for potential utilisation.

2. Materials and Methods (Preparation of Composites) and characterisation of basic properties

The materials used in this works are given in detail in table 1.The Various friction composite materials with varying ingredient is presented. Kenaf fibers were collected from Jeeva natural fibers, Chennai. The Composition of kenaf fiber contains 68.9% cellulose,0.3% lignin and 11.87% hemicellulose. Initially after collecting the ramie fibers it was chemically treated with NaOH solution having 3 weight percentages for a period of 24 hours. The Kenaf fibers were washed thoroughly with distilled water and dried for 6 hours at room temperature. And after that the kenaf fibers were cut down into length of 2-6mm length and used in frictional material composite as brake pad with varying percentages from 0 weight percentages to 15 weight percentages and designated as KFB1,KFB2,KFB3 and KFB4 and presented in table 1. With this the brake pads were fabricated as a frictional composites by following standardised procedure such as sequential mixing and followed by post curing. The specimen were cut into standard sized specimens of 25x25x6 millimetre cube. It is subjected to basic physical and mechanical test and friction test were carried out. The physio-Mechanical and chemical properties were analysed and average values were taken for 5 times and reported.Wensar density measuring kit was utilised to measure the density of the specimen.JISD4417 standard procedure were used to measure the porosity. Soxhlet extraction kit was used in measuring the acetone extraction for uncured resin.ASTMD570-98 was carried out in examining the ash content present in the composites-Scale intender was measured by using rockwell hardness tester kit by following the standards of D785.

Table1:Brake pad Composite Varying Ingredients(Materials Used)

		COMPOSITE DESIGNATIONS(Wt %)			
Sl.no	Ingredients	KFB1	KFB2	KFB3	KFB4
1	PF(Phenol Formaldehyde Resin)	15	15	15	15
2	Aramid Fiber	20	20	20	20
3	Natural Graphite	10	10	10	10
4	Vermiculates	10	10	10	10
5	Alumina	5	5	5	5
6	Barytes	40	35	30	25
7	Kenaf fibers	0	5	10	15

3. Results and Discussion

In this the results of brake pad composites is analysed for various characterisitcs.Various Physical, Chemical and Mechanical tests were done and it is presented in table 2. The Wensar density measuring kit was utilised to measure the density of the specimen. The specimen testing were carried out as pr industrial standards.

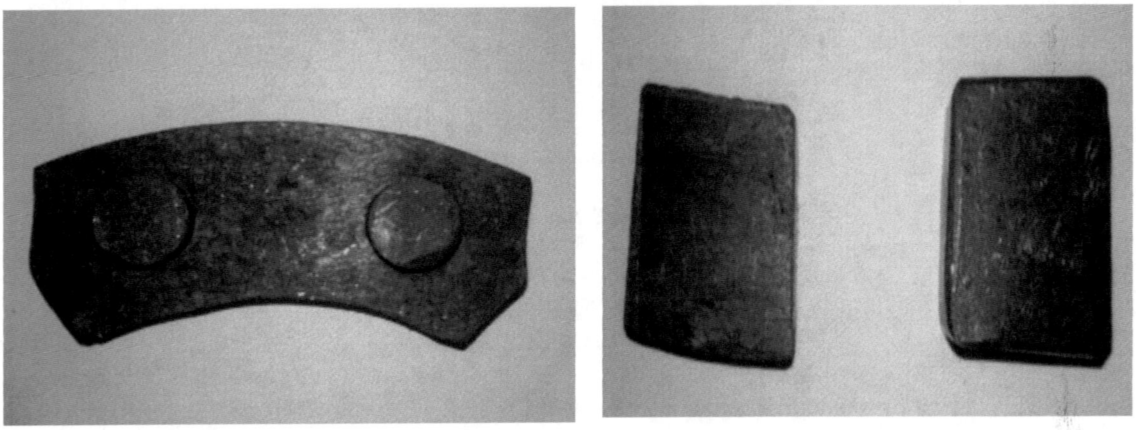

Figure1:Sample Images of Brake Pad

Table2:Friciton Composite Physical, Chemical and Mechanical Results.

		COMPOSITE DESIGNATIONS(Wt %)			
Sl.no	Examined Results	KFB1	KFB2	KFB3	KFB4
1	Density of the developed Specimen	2.52	2.46	2.34	2.24
2	Porosity.	7.08	7.68	8.39	9.28
3	Water Absorption Capacity	1.38	1.59	2.10	2.68
4	Ash Content in percentages	78.24	77.25	76.23	73.25
5	Acetone Extraction in Percentages	0.63	0.56	0.54	0.60

6	Hardness	112	108.6	107.64	105.23

3.1 Density of the developed composites:

Each composites were cut into standard specimen size and the density of the composites were measured. The density of composite KFB1 was 2.52,KFB2-2.46,KFB3-2.34 and KFB4-2.24.Density values got decreased by adding the kenaf fiber content. Declining trend was observed for the brake pad composites while calculating the density. It is mainly due to the properties of the fiber and the present of high density barytes.Due to the low weight percentage of kenaf and higher density of barytes the density values decreased. This is in tandem with the literature works done before[21].

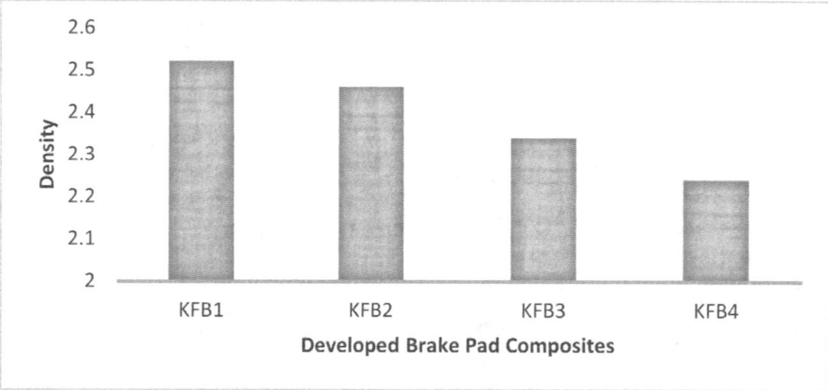

Figure 2:Density of developed Brake Pad Composites

3.2 Porosity and water absorption capacity of the developed composites

The void content(i.e porosity) shows an increasing trend values. Addition of fibers increased the percentages of void in the developed brake pad composites. In the present work also same trend was observed. Increase in the weight percentages of kenaf fiber content in the brake pad composites increased the porosity. The porosity of the developed composites KFB1-7.08,KFB2-7.68,KFB3-8.39 and for KFB4-9.28.In this also increase in the weight percentages of fibers increased the values of porosity. The main reason for this is because of structural integrity present in the fiber. Structural In-homogeneity present in the composites increased the void contents. The similar observations was seen in various work done before. Water absorption for all the brake pad composites were analysed and it was identified that the composites with KFB-1 possessed low water absorption capacity of 1.38 when compared to other. The decrease in the values of porosity also elucidated that the water absorption capacity and compressibility increase with the increase in the fiber weight percentages[22-24].

IOP Conf. Series: Materials Science and Engineering **961** (2020) 012016 doi:10.1088/1757-899X/961/1/012016

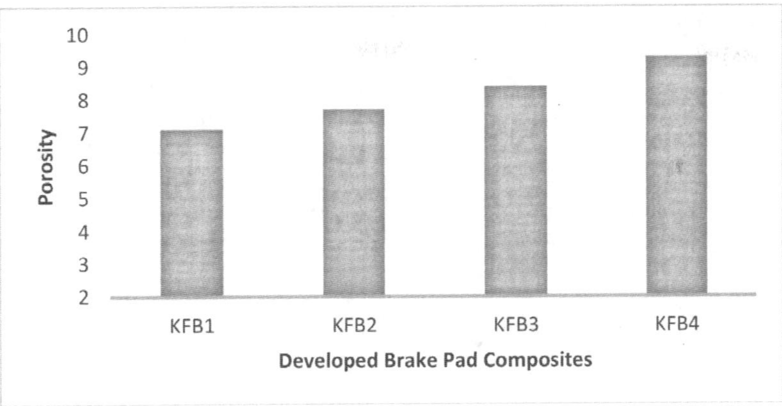

Figure 3:Porosity of developed Brake Pad Composites

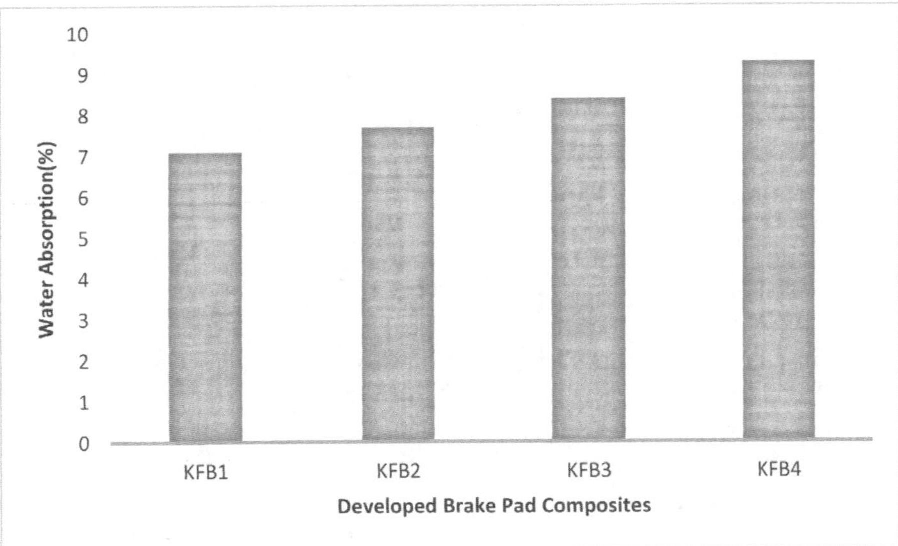

Figure 4:Water Absorption Capacity of the developed Brake Pad Composites

3.3 Hardness and Acetone Extraction of the developed Composites

The hardness of the developed brake pad composites is calculated by using hardness test rig with the Rockwell intender. The hardness of the developed composites in addition with the kenaf fibers possessed decreasing trend with the fiber loading at various percentages. The decrease in the hardness value is attributed owing to its inclination in the porosity of the brake pad composites. The Composites with 15 weight percentages had the lowest values of 105.23.Lower the porosity higher the hardness values were observed for the developed composites. The same trend were observed in various literatures done before[25].

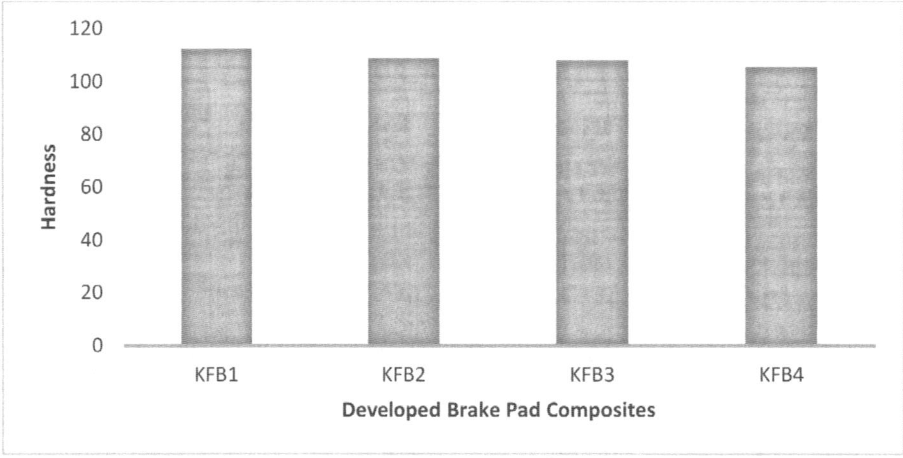

Figure 5:Hardness Value the developed Brake Pad Composites

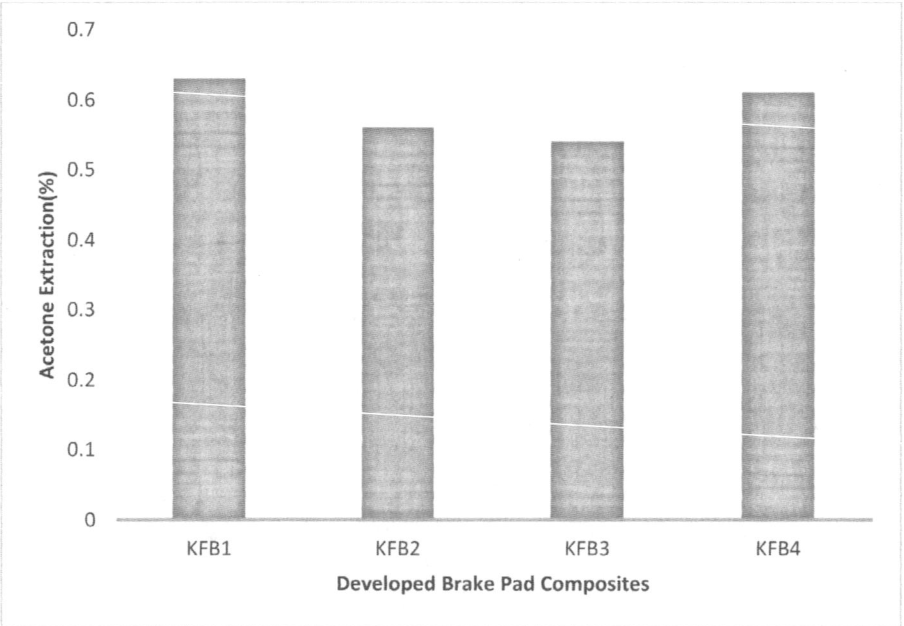

Figure 6:Acetone Extraction of the developed Brake Pad Composites

4. Conclusion

In present research, The brake pad friction material composites were developed by varying from 0 weight percentages to 15 weight percentages. The composites were fabricated by conventional technique and evaluated for various physio/chemical/mechanical properties. Based on the results it can be concluded that the kenaf fibers can be effectively used as an reinforcement material in frictional material applications. Density values got decreased by adding the kenaf fiber content. Declining trend was observed for the brake pad composites while calculating the density. hardness of the developed composites in addition with the kenaf fibers possessed decreasing trend with the fiber loading at various percentages. The decrease in the hardness value is attributed owing to its inclination in the porosity of the brake pad composites. . The porosity of the developed composites KFB1-7.08,KFB2-7.68,KFB3-8.39 and for KFB4-9.28.In this also increase in the weight percentages of fibers increased the values of porosity. The main reason for this is because of structural integrity present in the fiber

5.References

[1] Jafrey Daniel D, Sai Krishnan G and Velmurugan P Investigation on the characteristics of bamboo/jute reinforced hybrid epoxy polymer composites Mater. Res. Express 6 105346

[2] Sai Krishnan G, Jayakumari L S, Babu L G and Suresh G (2019) Investigation on the physical, mechanical and tribological properties of areca sheath fibers for brake pad applications Mater. Res. Express 6 085109

[3] C.Sivakandhan, Ganesh Babu Loganathan (2020) Material characterization and unconventional machining on synthesized Niobium metal matrix Mater. Res. Express 7 015018

[4] James J D, Manoharan S, Saikrishnan G and Arjun S (2019) Influence of bagasse/sisal fibre stacking sequence on the mechanical characteristics of hybrid-epoxy compositesJ. Nat. Fibers 2 1–11

[5] Saikrishnan G, Jayakumari L S, Vijay R and Lenin Singaravelu D (2019) Influence of iron–aluminum alloy on the tribological performance of non-asbestos brake friction materials–a solution for copper replacement Industrial Lubrication and Tribology 00 1–13

[6] Manoharan S, Shihab A I, Alemdar A S A, Ganesh Babu L, Vijay R and Lenin Singaravelu D 2019 Influence of recycled basalt-aramid fibres integration on the mechanical and thermal properties of brake friction composites Material Research Express 6 115310

[7] G Sai Krishnan, S Kumar (2020) Role of metal composite alloys in non-asbestos brake friction materials-A solution for copper replacement Materials Today: Proceedings 2020/3/31

[8] Vijay R and Singaravelu DL(2016). Experimental investigation on the mechanical properties of Cyperuspangorei fibers and jute fiber-based natural fiber composites. International Journal of PolymerAnalysisi and Characterization 2016; 21: 617–627

[9] Saikrishnan Ganesh Influence of Stacking Sequence on the Mechanical and Water Absorption Characteristics of Areca Sheath-palm Leaf Sheath Fibers Reinforced Epoxy Composites J Nat Fib 2020; 07: 2020

[10] Jothibasu S, Mohanamurugan S and Vijay R. Investigation on the mechanical behavior of areca sheath fibers/jute fibers/glass fabrics reinforced hybrid composite for lightweight applications. J Ind Tex 2020; 49: 1036–1060. DOI: 10.1177/1528083718804207

[11]. Chaudhary V, Bajpai PK and Maheshwari S. Studies on mechanical and morphological characterization of developed jute/hemp/flax reinforced hybrid composites for structural applications. J Nat Fib 2018; 15: 80–97

IOP Conf. Series: Materials Science and Engineering **961** (2020) 012016 doi:10.1088/1757-899X/961/1/012016

[12] Ganesh Babu, L. (2019). Influence of benzoyl chloride treatment on the tribological characteristics of Cyperus pangorei fibers based non-asbestos brake friction composites. *Materials Research Express*, *7*(1), 015303
[13] G.Sai Krishnan Babu L G and Pradhan R (2020) Study on tribological properties of palm kernel fiber for brake pad applications Materials Research Express Mater. Res. Express 7 015102
[14] Nongman. A. Fa, A. Baharina, and A.AbuBakarb ,—The Effect of Banana Leaves Lamination on the Mechanical Properties of Particle Board Panel‖ 5th International Conference on Recent Advances in Materials, Minerals and Environment (RAMM) & 2nd International Postgraduate Conference on Materials, Mineral and Polymer (MAMIP), 4-6 August 2015
[15] Ramnath, B. V., Kokan, S. J., Raja, R. N., Sathyanarayanan, R., Elanchezhian, C., Prasad, A. R., and Manickavasagam, V. M. (2013) Evaluation of mechanical properties of abaca–jute–glass fibre reinforced epoxy composite. Materials & Design, 51: 357-366.
[16] Vinod A, Vijay R and Singaravelu DL. (2018) Thermomechanical characterization of Calotropis gigantea stem powder-filled jute fiber-reinforced epoxy composites. J Nat Fibers Journal of Natural Fibers; 15: 648–657.
[17] G Sai Krishnan(2020) Investigation on the physical and mechanical characteristics of demostachya bipinnata reinforced with polyester composites Materials Today: Proceedings 2020/4/
[18] Tej Singh Sachin Tejyan Amar Patnaik(2019) Fabrication of waste bagasse fiber-reinforced epoxy composites: Study of physical, mechanical, and erosion propertiespolymer composites—2019
[19] Maha M. Ibrahima, Alain Dufresneb, Waleed K. ElZawawya,∗, Foster A. Agblevorc —Banana fibres and micro fibrils as lignocellulosic reinforcements in polymer composites‖ Article in Carbohydrate Polymers July 201
[20] Amira,*, Kamal AriffZainalAbidinb,FaizzatyBintiMdShiric —Effects of Fibre Configuration on Mechanical Properties of Banana Fibre/PP/MAPP Natural Fibre Reinforced Polymer Composite‖ Advances in Material & Processing Technologies Conference
[21] Singh T, Tejyan S, Patnaik A, Singh V, Zsoldos I and Fekete G 2019 Fabrication of waste bagasse fibre reinforced epoxy composites: study of physical, mechanical and erosion properties Polym. Compos. 40 3777–8
[22] G Sai Krishnan et al Investigation of Caryota urens fibers on physical, chemical, mechanical and tribological properties for brake pad applications Materials Research Express 015310 2019/12/16
[23] Singh T and Patnaik A 2017 Thermo-mechanical and tribological properties of multi-walled carbon nanotubes filled friction composite materials Polym. Compos. 38 1183–93
[24] Kumar V V and Kumaran S S 2019 Friction material composite: types of brake friction material formulations and effects of various ingredients on brake performance-a review Mater. Res. Express 6 082005
[25] Singh T, Tiwari A, Patnaik A, Chauhan R and Ali S 2017 Influence of wollastonite shape and amount on tribo-performance of nonasbestos organic brake friction composites Wear 386-387 157–64

AUTHOR INDEX